Biogas Systems in China

Bin Chen · Tasawar Hayat
Ahmed Alsaedi

Biogas Systems in China

Springer

Bin Chen
School of Environment
Beijing Normal University
Beijing
China

Ahmed Alsaedi
Faculty of Science
King Abdulaziz University
Jeddah
Saudi Arabia

Tasawar Hayat
Department of Mathematics
Quaid-i-Azam University
Islamabad
Pakistan

ISBN 978-3-662-55496-8 ISBN 978-3-662-55498-2 (eBook)
DOI 10.1007/978-3-662-55498-2

Library of Congress Control Number: 2017946673

Printed on acid-free paper

This Springer imprint is published by Springer Nature
The registered company is Springer-Verlag GmbH Germany
The registered company address is: Heidelberger Platz 3, 14197 Berlin, Germany

Preface

As the largest agricultural country in the world, almost 810 million tons of crop residues and more than 1600 million tons of animal manure could be collected for biomass resources every year in China, which has provided solid energy foundation for the daily life of rural residents (i.e., cooking and heating). However, with the economic reform since 1979, the energy structure of China's rural areas has experienced tremendous changes. The commercial energy began to replace the conventional biomass energy to be the dominant energy consumption categories. According to the National Bureau of Statistics of China, only 5% of agricultural residues was fully utilized, which leads to the disorder stacking and severe waste of agriculture slurry. In addition, the consumption amount of commercial energy was 1.5 times large as that in the 1990s, resulting in the risk of energy crisis and environmental degradation. Thus, it is urgent to provide adequate, clean and affordable energy (renewable energy) for rural residents to meet basic energy demand and raise living standards as well as the mitigation of carbon emissions.

Biogas as a renewable energy is well thought to be an important option to satisfy the growing energy demand of rural areas in developing countries. It can be generated by anaerobic digestion process using categories of locally available biomass residual, i.e., animal waste, domestic sewage, agricultural residue. The comprehensive utilization of biogas in the agricultural system is an important application for the development of recycling economy. This means the multiple reuse and recycling of biogas products and by-products produced substitute for chemical fertilizer, feed, additives as well as commercial energy of agricultural activities and household issues, through which a closed ecological chain with efficient materials flows and energy flows could be achieved, bringing about the multilevel benefits covering both economic characteristics and environmental performance.

With regard to the environmental performance, ecological service and support (nonmarketable energy, materials, ecosystem service for diluting the emissions and subsiding wastes) from the larger biosphere system are provided to guarantee upstream process demand of biogas system. Environmental cost for the resource generation and service support from a donor-side should be accounted to have a better understanding of the natural contribution to the biogas system. From the downstream perspective, although biogas and biogas by-products generated could be substitute for commercial energy and bring about the GHG mitigation, there exist nonrenewable energy, consumption and environmental emissions in the upstream of the biogas system, e.g., biogas digester construction, and emissions during the process of biogas fermentation and generation. It is therefore necessary to introduce quantitative measures to improve the understanding of the ecological and environmental performance covering both upstream and downstream processes of the biogas system. Apart from energy resource utilization and environmental impacts, the economic feasibility and efficiency should be investigated to find the hot spots of cost accounting. Currency outflows for initial construction and installation cost, maintenance cost as well as utilization cost should be considered. Moreover, both actual and potential monetary inflows for the substitute benefits of biogas products should be listed. It would be crucial to construct economic framework to have an integrated view of the monetary flows and find the key to improve economic feasibility.

The goal of this book is to evaluate the economic feasibility and efficiency, environmental impacts, renewability, and sustainability of the biogas system from different perspectives and diagnose the key sections in minimizing environmental impact and maximizing economic output. First, the conceptual biophysical economic framework with the nexus of economy–ecology–environment was constructed to evaluate eco-economic property of the biogas system considering both direct and indirect support and feedback between system and the external. Based on the accounting framework established, the economic cost–benefit analysis combined with data envelopment analysis (DEA) was conducted to assess the economic feasibility, and then, efficiency, optimization ways to improve the economic performance are proposed based on the DEA. A hybrid life-cycle assessment model was established to evaluate both direct and indirect GHG emission and reduction of the biogas project from a supply-side perspective, and possible pathways to achieve sustainable and low-emission discharge of biogas system were also analyzed based on scenario analysis. Finally, emergy thesis combined with financial accounting was used to recognize and quantify ecological service and ecosystem contribution in the upstream process from a donor-side perspective. Moreover, emergy-based statements and reports were provided to reflect environmental pressure, renewability, and sustainability, based on which managerial interventions can be made to achieve sustainable biogas system.

Beijing, China — Bin Chen
Islamabad, Pakistan — Tasawar Hayat
Jeddah, Saudi Arabia — Ahmed Alsaedi

Acknowledgements

The history of biogas production in China was reviewed by Bin Chen, Tasawar Hayat, and Ahmed Alsaedi. The main methods applied in the book were compiled by Bin Chen, Shaoqing Chen, Jin Yang, Binyue Zhang, Tasawar Hayat, and Ahmed Alsaedi. Four typical biogas systems in China were examined by Shaoqing Chen and Bin Chen. Environmental emissions from household biogas system were investigated by Bin Chen and Shaoqing Chen. Energy consumed by household biogas system was evaluated by Shaoqing Chen, Bin Chen, Tasawar Hayat, and Ahmed Alsaedi. The costs of household biogas project were investigated by Binyue Zhang, Bin Chen, Tasawar Hayat, and Ahmed Alsaedi. The emergy analysis used to evaluate the biogas-linked agricultural system is constructed by Shaoqing Chen, Jin Yang, and Bin Chen. The sustainability accounting and indicators system construction were conducted by Bin Chen, Jin Yang, Tasawar Hayat, and Ahmed Alsaedi.

We owe deep gratitude to the colleagues named above who led the work on individual chapters.

This work was also supported by the National Key Research and Development Program (2016YFA0602304), National Natural Science Foundation of China (No. 71573021, 71628301), special fund of State Key Joint Laboratory of Environment Simulation and Pollution Control (15L02ESPC).

Contents

1 History of Biogas Production in China 1
1.1 The History of Biogas in China 4
1.2 Significance of Developing Biogas in China 4
1.3 Policies for Biogas Projects 5
1.3.1 Energy Policies for Biogas Projects 5
1.3.2 Environmental Policies for Biogas Projects 6
1.3.3 Economic Policies for Biogas Projects 7
1.3.4 Standards for Biogas Projects 8
1.4 Development of Biogas Industrialization 10
1.4.1 Introduction of Biogas Production 10
1.4.2 Biogas Plant Industry 12
1.4.3 Biogas Equipment Industry 12
1.4.4 Biogas Service Industry 12
1.5 Development Planning for Biogas Production 13
1.5.1 Developing Motion of Biogas Production 13
1.5.2 Rural Biogas Development Policies 14
References 14

2 Main Methods 17
2.1 Life-Cycle Assessment (LCA) 17
2.1.1 Background of LCA 17
2.1.2 Methodology of LCA 19
2.1.3 LCA-Based Integrated Evaluation Indices 20
2.2 Economic Assessment 22
2.2.1 Cost–Benefit Analysis 22
2.2.2 Data Envelopment Analysis (DEA) 23
2.3 Emergy Analysis 25

2.3.1 Emergy Concept 25
2.3.2 Emergy Diagram 26
2.3.3 Emergy Indices 26
2.3.4 Emergetic Ternary Diagrams 29
2.4 Extended Exergy Analysis 30
2.4.1 Extended Exergy Analysis Framework 30
2.4.2 Extended Exergy-Based Sustainability Indexes 31
2.5 Analytic Hierarchy Process 34
References 34

3 Four Typical Biogas Systems in China 37
3.1 "Six-in-One" Biogas System (SIOBS) 37
3.1.1 Biogas Plant Construction Stage 38
3.1.2 Biogas Plant Maintenance Stage 38
3.1.3 Feedstock Supply Stage 39
3.1.4 Biogas Energy Utilization Stage 39
3.1.5 Digestate Processing Stage 39
3.2 Biogas-Persimmon Cultivation and Processing System (BCPS) . . . 40
3.2.1 Biogas Infrastructure (Fraction I) 41
3.2.2 Biogas System Operation (Fraction II) 41
3.2.3 Digestate Reuse for Persimmon Cultivation (Fraction III) 42
3.2.4 Biogas Utilization for Persimmon Processing (Fraction IV) 43
3.3 Wastewater Treatment Plants (WWTPs) 44
3.3.1 WWTP Construction Stage 44
3.3.2 WWTP Operation and Maintenance Stage 44
3.3.3 Biogas–Sludge Use Stage 45
3.4 "Three-in-One" Biogas Project 45
References 47

4 Environment Emissions of Household Biogas Project 49
4.1 Biogas-Persimmon Cultivation and Processing System (BCPS) 49
4.1.1 Emissions Inventory 49
4.1.2 GHG Emissions Mitigation 51
4.2 Wastewater Treatment Plants (WWTPs) 52
4.2.1 Emissions Inventory 52
4.2.2 Environmental Impact Mitigation 52
4.3 'Six in One' Biogas System (SIOBS) 56
4.3.1 System Inventory 57
4.3.2 Overall Environmental Evaluation 58
4.3.3 Sensitivity Analysis 59
4.3.4 Digestate Reuse 60

4.3.5 Environmental Benefit Under Different Scenarios 62
References 63

5 Energy Evaluation of Household Biogas Project 65
5.1 Life-Cycle Energy Evaluation 65
5.2 Case Study of BCPS 66
5.3 "Six in One" Biogas System (SIOBS) 68
5.3.1 Overall Energy Evaluation 68
5.3.2 Sensitivity Analysis 68
5.3.3 Environmental Benefit Under Different Scenarios 69
5.3.4 Digestate Reuse 70
5.4 Wastewater Treatment Plants (WWTPs) 70
5.4.1 Energy Input–Output Analysis 70
5.4.2 System Energy Production 70
5.4.3 Sensitivity Analysis 72
References 73

6 Economic Assessment of Household Biogas Project 75
6.1 Economic Feasibility Analysis 76
6.1.1 System Boundary 76
6.1.2 Economic Costs and Benefits 76
6.1.3 Financial Analysis 77
6.2 DEA Analysis 79
6.2.1 Input and Output Variables 79
6.2.2 Economic Efficiency Evaluation 80
6.2.3 Economic Efficiency Optimization 81
6.3 Policy Implication 82
References 83

7 Emergy Analysis of Biogas-Linked Agricultural System 85
7.1 Emergy Analysis 85
7.1.1 Methodology 86
7.1.2 Results and Discussion 90
7.1.3 Concluding Remarks 98
7.2 Scenarios Analysis Based on Emergy Analysis 99
7.2.1 Materials and Methods 99
7.2.2 Emergy-Based Analysis of BLAS 105
7.2.3 Scenario Analysis 106
7.2.4 Discussion and Conclusions 108
7.3 Three-Level Emergetic Evaluation 110
7.3.1 Materials and Methods 112
7.3.2 Results and Discussion 115
7.3.3 Conclusions 134
References 134

8 Sustainability and Indicator System . . . 137
8.1 Extended Exergy-Based Sustainability Accounting . . . 137
8.1.1 System Boundary . . . 138
8.1.2 Results and Discussion . . . 138
8.1.3 Conclusions . . . 144
8.2 Indicator System for Impact Assessment . . . 146
8.2.1 Data Sources . . . 146
8.2.2 Analytic Hierarchic Process . . . 148
8.2.3 Results and Discussion . . . 149
8.2.4 Conclusions . . . 151
References . . . 151

About the Authors

Prof. Bin Chen obtained his Ph.D. in Environmental Science from Peking University and did post-doctoral study in Harvard Kennedy School. He is currently Full Professor of Beijing Normal University. Dr. Chen's research focuses on the urban metabolism and nexus issues. He has published over 250 peer-reviewed papers in prestigious international journals such as PNAS, Trends in Biotechnology, and Nature Climate Change. His works are widely recognized and have more than 5000 citations with H-index of 39 according to Web of Science. He is serving as Editor-in-Chief of Energy, Ecology and Environment, Associate Editor of Journal of Cleaner Production, Frontiers of Earth Science, Subject Editor of Applied Energy, and an editorial board member of Ecological Modelling, Journal of Environmental Management, Journal of Hydrodynamics and Ecological Informatics, etc.

Prof. Tasawar Hayat (born in Khanewal, Punjab), Distinguished National Professor and Chairperson of Mathematics Department at Quaid-I-Azam University Islamabad, Pakistan, is renowned worldwide for his seminal, diversified and fundamental contributions in models relevant to mathematical modeling, physiological systems, control engineering, climate change, renewable energy, low-carbon technologies, environmental issues, non-Newtonian fluids, wave mechanics, homotopic solutions, stability, nanomaterials, and in several other areas. He has an honor of being fellow of

Pakistan Academy of Sciences, Third World Academy of Sciences (TWAS) and Islamic World Academy of Sciences in the mathematical Sciences. He is the HIGHLY CITED RESEARCHER from Pakistan in both MATHEMATICS and ENGINEERING. His national and international recognition is evident by the membership of international and national Committees, leadership and motivation, numerous scholarships and fellowships, conducted research projects, convened many national and international conferences, seminars delivered and attended conferences the world over, established research collaboration with leading international scientists, associate editor/editorial membership of the international journals including ISI, reviewer of the international journals and MS and PhD students produced. His publications in diverse areas are in high-impact-factor journals. His research work has total ISI WEB citations (over 20000) and h-index (62) at present. His citations through google scholar are over 30000 and h-index more than 80. He has over 1000 SCI journal publications at present. He has produced 37 PhD and 115 MS students. At present 20 (12 PhD and 8 MS) students are working under his supervision. He has received many national and international awards including Tamgha-i-Imtiaz, Sitara-i-Imtiaz, Khwarizmi Int. award, ISESCO Int. award, COMSTECH Int. Award, TWAS prize for young scientists, Alexander-Von-Humboldt fellowship, etc.

Prof. Ahmed Alsaedi obtained his Ph.D. degree from Swansea University (UK) in 2002. He has a broad experience of research in applied mathematics. His fields of interest include dynamical systems, nonlinear analysis involving ordinary differential equations, fractional differential equations, boundary value problems, mathematical modeling, biomathematics, Newtonian and non-Newtonian fluid mechanics. He has published several articles in peer-reviewed journals. He has supervised several MS students and executed many research projects successfully. He is reviewer of several international journals. He served as the chairman of the mathematics department at KAU, and presently he is serving as director of the research program at KAU. Under his great leadership, this program is running quite successfully and it has attracted a large number of highly rated researchers and distinguished professors from all over the world. He is also the head of NAAM international research group at KAU.

List of Figures

Fig. 1.1 Development of biogas projects in China 2
Fig. 1.2 Strategic development mode of biogas production in China . . . 11
Fig. 2.1 LCA assessment framework. 20
Fig. 2.2 Emergy systems language 27
Fig. 2.3 Properties of emergetic ternary diagrams 30
Fig. 3.1 Gongcheng county located in Guangxi province, China 38
Fig. 3.2 "Six-in-One" biogas system (SIOBS) 38
Fig. 3.3 Overview of a household biogas project coupled to persimmon cultivation and processing (*FI* Fraction I, *FII* Fraction II; *FIII* Fraction III; *FIV* Fraction IV). Reprinted from (Chen and Chen 2013), with permission from Elsevier 41
Fig. 3.4 "Three-in-One" biogas production system. Reprinted from (Yang and Chen 2014), with permission from Elsevier 45
Fig. 3.5 Scheme of the evaluation of the household biogas project. Reprinted from (Yang and Chen 2014), with permission from Elsevier 46
Fig. 4.1 Processes contribution to GHG emission mitigation in *1* Fraction I, *2* Fraction II, *3* Fraction III, and *4* Fraction IV. Reprinted from (Chen and Chen 2013), with permission from Elsevier 51
Fig. 4.2 Proportional effects of different stages in environmental impact mitigation by a WWTP for scenarios: *1* WS0, *2* WS1, *3* WS2, *4* WS3, *5* WS4, *6* WS5, *7* WS6, and *8* WS7 55
Fig. 4.3 System boundary of the study 57
Fig. 4.4 Contribution of different stages to the environmental condition improvement 59
Fig. 4.5 Sensitivity analysis of GHGs mitigation 60
Fig. 4.6 Processes contribution to environmental condition improvement within digestate processing stage 61
Fig. 4.7 Environmental condition improvements under different digestate reuse scenarios 62

Fig. 5.1 Processes' contribution to energy production in *1* Fraction I, *2* Fraction II, *3* Fraction III, and *4* Fraction IV. Reprinted from (Chen and Chen 2013a), with permission from Elsevier 67
Fig. 5.2 Sensitivity analysis of NEP mitigation. 69
Fig. 5.3 Energy consumption and production during each stage of WWTP under different scenarios 71
Fig. 5.4 Sensitivity analysis of WWTP energy production. Reprinted from (Chen and Chen 2013b), with permission from Elsevier. . 72
Fig. 6.1 System boundary of biogas project 76
Fig. 6.2 Substitute economic benefit structure of household biogas project 77
Fig. 6.3 Net present cost and benefit (economic value) of the biogas project from 2009 to 2018. 79
Fig. 6.4 Economic efficiency of biogas projects 80
Fig. 7.1 Emergy diagram of the BLAS. Reprinted from Yang and Chen (2014), with permission from Elsevier 87
Fig. 7.2 Emergy inputs of the BLAS 90
Fig. 7.3 Comparisons of emergy indices of different agricultural systems (*1* BLAS in this paper, *2* Chinese agricultural system in 2000, *3* Italian agricultural system in 2000, *4* a "Four-in-One" peach production system in Beijing) 91
Fig. 7.4 Aggregated emergy flow diagram for BLAS. Reprint from Chen and Chen (2012), with permission from Elsevier 100
Fig. 7.5 Emergetic ternary diagrams for *1* system aggregation analysis, *2* input sensitivity analysis, *3* resource equivalence analysis, and *4* sustainability zone analysis. Reprinted from Chen and Chen (2012), with permission from Elsevier 101
Fig. 7.6 Emergetic ternary diagrams for *1* BLAS and *2* its subsystems. *Notes Point a* planting subsystem; *Point b* breeding subsystem; *Point c* aquaculture subsystem; *Point d* biogas subsystem; *Point S* simergic point of all subsystems (also the BLAS's point calculating the overall system). *F*-sensitivity line is a straight line joining an apex (*F*) to point 1. Reprinted from Chen and Chen (2012), with permission from Elsevier . . 108
Fig. 7.7 Emergetic ternary diagrams of BLAS under five scenarios for *1* planting subsystem, *2* breeding subsystem, *3* aquaculture system, *4* biogas subsystem, and *5* overall BLAS. *Note* The size of system points is proportional to their total emergy input values. Reprinted from Chen and Chen (2012), with permission from Elsevier. 111
Fig. 7.8 Regression relations between EMI and SI (*1*), EMI and BEE (*2*), and BEE and SI (*3*) for current status (represented as "O") and five scenarios of BLAS. Reprinted from Chen and Chen (2012), with permission from Elsevier. 112

Fig. 7.9 Conceptual framework of three-level emergetic evaluation. *Note* RC stands for resource component. Reprinted from Chen and Chen (2014), with permission from Elsevier. 113
Fig. 7.10 Aggregated emergy flow diagram of a BLAS. *Notes RR* renewable resources; *NR* nonrenewable resources; *NP* nonrenewable purchases; *NP* nonrenewable purchases. Reprinted from Chen and Chen (2014), with permission from Elsevier 114
Fig. 7.11 ECRs of different resources with each production process of the BLAS over time: *1* planting production, *2* aquaculture production, *3* breeding production, *4* biogas production, and *5* overall production process. Reprinted from Chen and Chen (2014), with permission from Elsevier. 132
Fig. 7.12 ESEs of different resources within each production process of the BLAS over time: *1* planting production, *2* aquaculture production, *3* breeding production, *4* biogas production overall production process. "S line" stands for average level (which equals 1.00), and *5* overall production process. Reprinted from Chen and Chen (2014), with permission from Elsevier 133
Fig. 8.1 Cumulative exergy inputs of the biogas project. 140
Fig. 8.2 Lifetime greenhouse gas emissions of the biogas project. 142

List of Tables

Table 1.1	Policies, laws, rules for the household biogas projects	3
Table 1.2	Standards for biogas	9
Table 2.1	LCA-based integrated evaluation indices	22
Table 2.2	Indicators for the energy and emergy analysis	28
Table 2.3	Evaluation indicator framework for biogas project	33
Table 4.1	Aggregated annual inventory of coupling biogas production to persimmon cultivation and processing[a]	50
Table 4.2	Annual environmental emissions inventory of a WWTP	53
Table 4.3	Environmental impact mitigation by WWTP under different scenarios	54
Table 4.4	Life-cycle emissions inventory of a SIOBS (on per year basis)	58
Table 4.5	Improvement of environmental performance by SIOBS	59
Table 4.6	GHGs mitigation within digestate processing stage	61
Table 5.1	Energy inventory of coupling biogas production to persimmon cultivation and processing	66
Table 5.2	Improvement of energy performance by SIOBS	68
Table 6.1	Input and output inventory of the household biogas project	78
Table 6.2	Selection of input and output variables (one year)	80
Table 6.3	Optimization of input and output variables of biogas projects	81
Table 7.1	Emergy-based indices and ratios for the BLAS	89
Table 7.2	Selected results of emergy analysis of the BLAS	91
Table 7.3	Emergy analysis of planting subsystem	93
Table 7.4	Emergy analysis of breeding subsystem	95
Table 7.5	Emergy inputs and outputs of biogas subsystem	96
Table 7.6	Carbon emission reduction potential of each biogas digester per year in Gongcheng	97
Table 7.7	Comparison of CO_2 emission indicators for different biomass utility alternatives	97
Table 7.8	Comparison of emergy-based indices of three subsystems	98

Table 7.9 Emergy-based indices for BLAS evaluation (Brown and Ulgiati 2004a, b; Odum 1996; Ulgiati and Brown 1998). 102
Table 7.10 Descriptions of five developmental scenarios of BLAS 103
Table 7.11 Emergy evaluation of overall BLAS 104
Table 7.12 Emergy evaluation of BLAS's subsystems 107
Table 7.13 Emergy-based indices of BLAS and its subsystems. 108
Table 7.14 Aggregated emergy evaluation of BLAS under different scenarios 109
Table 7.15 Analytical table for emergetic evaluation at levels II and III 115
Table 7.16 Emergy accounting of a BLAS in Gongcheng County (2008) 116
Table 7.17 Emergy accounting of the BLAS in Gongcheng County (2004) 118
Table 7.18 Emergy accounting of the BLAS in Gongcheng County (2000) 120
Table 7.19 Decomposed emergy flows within various production processes of the BLAS (year 2000) (*unit* sej) 122
Table 7.20 Decomposed emergy flows within various production processes of the BLAS (year 2004) (*unit* sej) 123
Table 7.21 Decomposed emergy flows within various production processes of the BLAS (year 2008) (*unit* sej) 124
Table 7.22 Decomposed emergy indices within various production processes of the BLAS from 2000 to 2008 125
Table 7.23 Emergy contribution rates (ECRs) of different resources within each production process of the BLAS (year 2000) 126
Table 7.24 Emergy supply efficiencies (ESEs) of different resources within each production process of the BLAS (year 2000) 127
Table 7.25 Emergy contribution rates (ECRs) of different resources within each production process of the BLAS (year 2004) 128
Table 7.26 Emergy supply efficiencies (ESEs) of different resources within each production process of the BLAS (year 2004) 129
Table 7.27 Emergy contribution rates (ECRs) of different resources within each production process of the BLAS (year 2008) 130
Table 7.28 Emergy supply efficiencies (ESEs) of different resources within each production process of the BLAS (year 2008) 131
Table 8.1 Inventory analysis of the cumulative exergy demand of a household biogas digester. 139
Table 8.2 Inventory of the economic inputs of a household biogas digester. 141
Table 8.3 GHG emission reduction potential of each biogas digester per year 143

Table 8.4 Comparison of different renewable energy utilization projects with extended exergy-based sustainability indicators . . . 143
Table 8.5 Five-grade classification . . . 148
Table 8.6 Impacts of biogas project on the sustainable development of agro-ecosystem . . . 149
Table 8.7 Environmental, social, and economic benefits derived from biogas project . . . 150

Chapter 1
History of Biogas Production in China

Abstract The biogas digestion has experienced rapid development for its unique characteristics for dealing with energy shortage and environmental pollution. In this chapter, we reviewed the development trend of biogas projects in China from 2000 to 2014. The significance of developing biogas in China was analyzed by accounting the energy supply and emission reduction. The details about the biogas industrialization were then probed into, with the conclusion that the smallholder farms are gradually transforming to larger-scale farms. The strategic development mode of biogas production industry was summarized. Then, the relationship among three types of biogas industries including biogas plant industry, biogas equipment industry, and biogas service industry was analyzed in detail. Finally, the development planning and development motion for biogas production as well as corresponding policies in China were reviewed. The policies, including energy policies, environmental policies, and economic policies, laws, rules and regulations, and standards for the household biogas projects were summarized, which may contribute to the rapid promotion of biogas project.

Keywords Biogas industrialization · Biogas supply and demand · Scales of biogas project · Biogas plant industry · Biogas equipment industry · Biogas service industry

Overview

There is a long history for the utilization of biogas in China (Feng et al. 2012). In the 1880s, the first try of modern biogas digester was built in Guangdong. Between 1900s and 1990s, the biogas digesters have undergone two transformations with the increasing demand for sanitation improvement. Since the 1990s, the biogas digestion has experienced rapid development for its characteristics for dealing with energy shortage and environmental pollution (Deng et al. 2017). So far, household

B. Chen et al., *Biogas Systems in China*, DOI 10.1007/978-3-662-55498-2_1

biogas digester ("Three-in-One," "Four-in-One," and "Five-in-One" modes) and medium- and large-scale biogas plants have been developed as the main categories of biogas systems in China (Zhang and Chen 2016). Figure 1.1 shows the developing trend of the biogas digesters mentioned above of China from 2000 to 2014. The substantial development of biogas is closely linked to both the geographic condition and the incentive of policy from government. With regard to the geographic advantages, various fermentation materials can be gained annually. As estimated, the production amount of livestock manure could reach 4 billion tons in 2020 and also almost 550 million tons of residue wastes could be gained for fermentation process (MOA 2007).

There are mainly two sources from the governmental support: financial subsidies and policy incentives. From 2001 to 2012, 34.6 billion CNY (abbreviation of Chinese Yuan) from the government funding was invested in the development of biogas projects from central government funds for biogas development, nearly 75% of which was provided for the rural resident (Wang et al. 2016). Apart from the financial support, a series of government policies, laws, and regulations were also promulgated. Table 1.1 has shown the main policies, laws, and regulations for the biogas of China, which has made large contribution to the rapid promotion of biogas project. Due to the fact that biogas projects develop fast in China rural areas, it is of great importance to establish an accounting framework to have a systematic and objective assessment on the comprehensive performance of biogas project.

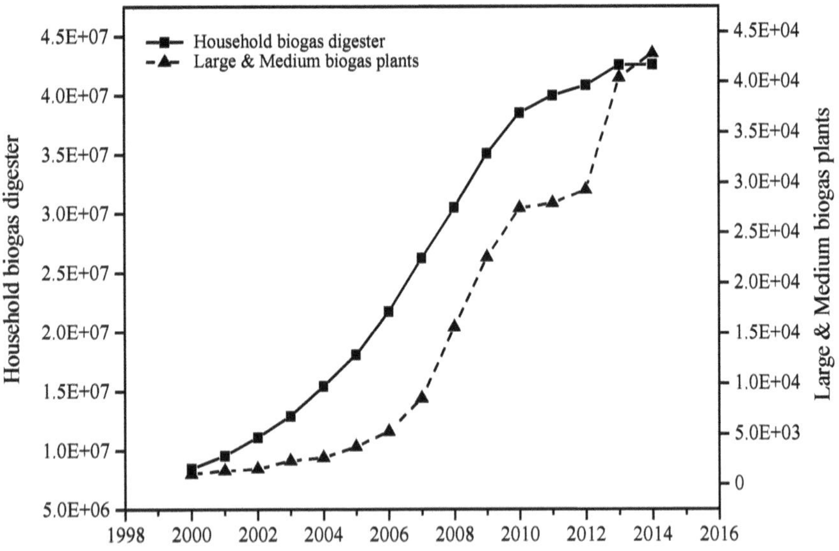

Fig. 1.1 Development of biogas projects in China

Table 1.1 Policies, laws, rules for the household biogas projects

	Period/Code	Events
Policies	2000	Biologically Enrichment of Rural Projects
	2003	Debt Project on Rural Household Biogas
	2006	National Countryside Biogas Construction Plan (from 2006 to 2010)
	2007	Rural Biogas Projects
	2008	Act on the Circular Economy Development
	2012	The 12th Five-Year Plan for Bioenergy Development(2011–2015)
Laws	2002	Agricultural Law
	2005	Renewable Energy Law
	2005	Animal Husbandry Law
	2007	Energy Conservation Law
	2008	Circular Economy Development Act
	2009	Renewable Energy Law Revision
Rules and regulations		Administration Rules for Fund of Rural Biogas Construction Project
		Scheme of Rural Biogas Service System
		Rural Resident Community Plan (related households biogas construction)
Standards	GB/T 4750-2002	Collection of Standard Layout for Household Biogas Projects
	GB/T 4751-2002	Standard for Examination of Quality for Household Biogas Projects
	GB/T 4752-2002	Operation Rules for Construction of Household Anaerobic Digesters
	GB/T 26715-2011	Biogas Clique Standard
	NY/T 1639-2008	Technical Specification for Rural Biogas Projects and Three Renovations
	NY/T 1704-2009	Regulation of Biogas Power Generation
	NY/T 1916-2010	Criterion for Facilities for Biogas Digestate
	NY/T 667-2011	Scale Classification for Biogas Plants
	NY/T 2065-2011	Technical Standard for Digestion Fertilizer Utilization

Reprinted from (Wang et al. 2016), with permission from Elsevier

1.1 The History of Biogas in China

China has recorded the history of biogas utilization long time ago. Far in the first century B.C., the first human gas well, named Lin Qiong fire well, was drilled, and then, other wells were drilled. During the period of the Warring Kingdoms (475–221 B.C.), an administrator called Li Bing supervised the production of natural gas. In 1980, there were many pilots on making gas in Guangdong Province. At the end of the nineteenth century, the biogas production mode was initiated after the tank was invented.

In the 1920s, Guorui Luo in Taiwan Province created the hydraulic biogas digesters and established the first institution for promoting gas in Shantou city of Guangdong Province. In 1931, he moved to Shanghai, founded the head office of commercial gas, and developed more than 10 branches that spread the biogas utilization to 13 provinces.

In 1958, the second peak of biogas development appeared, and biogas plants were built in many areas across the nation. However, lack of technology and management skills, few biogas plants were put into final use. Entering the 1970s, due to shortage of rural fuels, provinces, such as Sichuan, Jiangsu, Henan, developed biogas very quickly. In a few years, the total number of biogas plants had increased to over 70 million. Due to unreasonable production and raw methods, the service life of built biogas plants was about 3–5 years, and after that, most biogas plants were not workable. Later in the 1980s, the State Council set up the National Biogas Construction Steering Panel to summarize the experiences and lesson of biogas practice. Furthermore, Chinese Association of Biogas coordinated over 1700 technical workers to conduct basic research and introduce advanced anaerobic technologies abroad, proposing the construction policies like “Suiting Measures to Local Conditions, Adhering to Quality, Constructing and Management Equally Important, Comprehensive Utilization, Pursuing Pragmatism, Positive and Steady Development.” The anaerobic research, including a set of scientific technologies of plant building for rural household hydraulic biogas digesters, fermentation technology, and supporting equipment, was boosted. Thus, the strengthened management, promotion of production mode, and quality improvement have promoted the biogas production to a new stage of healthy and stable development.

1.2 Significance of Developing Biogas in China

China is undergoing rapid urbanization and industrialization consuming a huge amount of energy resources, which has become the largest energy consumer in the world. In 2015, the total energy consumption reached 4.3 billion t of standard coal equivalent (tce), of which about 64% was coal consumption and net imported natural gas reached 62.1 billion m^3. In order to achieve the targets of G20 Summit and Paris World Climate Change Conference, nonfossil energy proportion of primary energy consumption has to increase to 20%. It is estimated that the construction of one biogas production plant with the productivity of 0.1 million cubic meters can generate 3.65 million cubic meters, i.e., 4343 tce, which may reduce the primary energy consumption.

Simultaneously, Chinese government encourage people from rural areas to move to the nearest urban regions or small- and medium-sized cities to accelerate urbanization. This trend may increase the energy demand, but the existing urban gas supply system is difficult to cover the new urbanization area for the high cost of pipe laying and transportation. It is thus of great strategic significance to develop biogas to increase the supply of clean energy and reduce environmental emission reduction to promote sustainable development.

In addition, there are huge amount of agricultural waste resources in China, which include 1.04 billion tons of crop straws and 2.05 billion livestock and poultry manure per year. The potential quantity of crop straw is around 0.9 billion, of which 0.18 billion crop straw has not been utilized effectively. Previously, rural household used to burn straw as energy source, animal feed, and fertilizer (Hu 2008). However, many commercial energy companies began to supply energy in rural areas, changing the rural energy consumption structure in China over time, leading to significant decline of burning crop straw as fuel (Pachauri and Jiang 2008). These wastes thus were burnt at random, discarded, or directly discharged into the environment, which may cause much hazardous environmental pollution (Sun et al. 2005). Similarly, there are 56% of livestock and poultry manure under insufficient use. It is estimated that the construction of large-scale biogas project with the pool capacity of 5000 cubic meters can consume 30,000 tons of manure or 0.6 million tons' dry straw, reducing emission of 1500 tons of Chemical Oxygen Demand (COD) or 90 tons of particulate matter.

1.3 Policies for Biogas Projects

The Chinese government strongly supports the development and application of biogas energy to optimize energy structure and solve the problem of rural energy supply. Given this, fruitful polices and project have been formulated to promote the development of biogas (Zhang et al. 2009).

1.3.1 Energy Policies for Biogas Projects

In 1986, the China National Economic Committee issued the first policy paper on renewable energy development, titled "Circular on Improving Rural Energy

Development", which emphasized the importance of bioenergy. During the Eighth Five-Year Plan (1991–1995) period, the government placed the strategic measure of renewable energy in a high position for achieving goals in "Agenda 21" as well as national economic development targets. In 1995, Chinese government approved the "1996–2010 New Energy and Renewable Energy Development Programme", which promulgated the development of renewable energy (Zhao and Fan 2010).

"Renewable Energy Law" was released in 2005 and came into force on January 1, 2006. The "Renewable Energy Prices and Cost-sharing Management Trial Procedures" was also issued to encourage domestic companies to boost renewable energy development. This regulation pointed out that the price of power generated by biomass should be decided by the central government and the subsidy standard of power was set at 0.25 CNY per kWh based on desulphurized coal electricity price transferred into national grid (National Development and Reform Committee of China 2006). "Regulations Related to Renewable Energy Power Generation", which was released by the National Development and Reform Committee, clarified the approval and management methods of renewable energy power generation projects to support regulations for "Renewable Energy Law" and the "Renewable Energy Prices and Cost-sharing Management Trial Procedures". This regulation also provided guidance for the implementation of standards for enterprises entering the renewable energy sector and encouraged domestic enterprises to become involved in the development of renewable energy (National Development and Reform Committee of China 2006). Continually, Chinese government published the "Medium and Long-term Development Program for Renewable Energy" in 2007. On April 1, 2010, the revised "Renewable Energy Law" became effective, which stipulated that the country would implement a fully supported system for the power generated by renewable energy (National People's Congress of the People's Republic of China 2009). Therefore, renewable energy has been given the priority in China.

1.3.2 Environmental Policies for Biogas Projects

In past decades, Chinese government focused on environment development and released a series of environmental laws, regulations, and standards, which were adopted to support environmental protection and to develop bioenergy production. Some important regulations include "Environmental Protection Law", "Water Pollution Prevention Law", "Atmospheric Pollution Prevention Law", "Integrated Emission Standard of Air Pollutants" (GB16297-1996)," Standards for Irrigation Water Quality" (GB5084-92)", "Quality Standard for Ground Water" (GB/T14848-93), and "Environmental Quality Standards for Surface Water" (GB3838-2002).

To deepen the environmental protection, China established the "Environmental Impact Assessment System" and the "Three Simultaneous Systems", indicating that the factory and facilities for waste treatment should be designed, constructed, and opened at the same time to facilitate environmental management of construction projects through the "Environmental Impact Assessment Law" and "Environmental Protection Management Regulations for Construction Projects". These regulations forced construction projects to implement an environmental impact assessment and required construction projects to design, construct, and implement environmental protection (Ministry of Environmental Protection of China 2010). In terms of controlling the pollution from livestock and poultry production facilities, the following measures have been established and implemented (Ministry of Environmental Protection of China 2010): "Discharge Standard of Pollutants for Livestock and Poultry Breeding" (GB18569-2001), "Management Approach for Pollution Prevention of Livestock and Poultry Farms" (2001), "Criteria for Evaluating the Environmental Quality of the Livestock and Poultry Farm" (GB/T19525.2-2004), and "Technical Specifications for Pollution Treatment Projects of Livestock and Poultry Farms" (HJ497-2009).The construction of medium- and large-scale livestock and poultry farms is subject to the "Environmental Impact Assessment System" and the "Three Simultaneous Systems". Biogas technology is widely recognized as an environmentally friendly measure, except the energy production phase, which may contribute to a more efficient and safe recycling of manure to fields or fisher ponds (Chen et al. 2010).

1.3.3 Economic Policies for Biogas Projects

Since the beginning of the 1970s, the Chinese government has declared to strongly support the construction of biogas digesters on livestock farms by providing energy loans with an interest-free system. The financial supports target at large- and medium-scale biogas plant projects, solar energy utilization, and wind power generation technology (Gao et al. 2006).

Biogas construction projects have been widely implemented all over the country since the "Rural Ecological Enrichment Project" was proposed by the Ministry of Agriculture. The government has motivated the innovations of environmentally friendly energy and technologies, especially biogas digesters and thus has given high priority to the rural biogas sector through many supportive projects, such as rural small-scale, public, infrastructure projects and rural basic construction projects since 2003 (Ministry of Agriculture and National Development and Reform Commission of China 2004).The total investment from central government to the rural biogas industry has reached over 19.0 billion CNY, of which the construction of household biogas digester accounted for about 82%, the construction of medium- and large-scale biogas plants accounted for about 10%, and the financed service systems accounted for about 8% from 2003 to the end of 2009. Referring to the "Management Approach of National Debt Project for the Rural Biogas Construction (Trial)" (Ministry of Agriculture and National Development and Reform Commission of China 2004), it is declared that the subsidy provided by central government ranged from 800 CNY to 1200 CNY per 8 m^3 of household biogas digester in different regions and 20,000 CNY per 100 m^3 of large- and medium-scale biogas digester for construction materials such as wages of special equipment, cement, pebbles and bricks, but the domestic users were required to pay the remaining costs. Therefore, the effective investment at the beginning of twenty-first century led to rapid growth of biogas sector every year in China (Cao 2009). In 2010, the renewable energy development was pushed by the revised "Renewable Energy Law", consisting of state special fund and additional renewable energy surcharge 0.002 CNY per kWh in the sale price of power. This fund provides financial support to the power-generating plants through the provision of a premium price for the power originating from renewable energy production (NPC 2009).

To mitigate the greenhouse gas emissions and global warming, the Clean Development Mechanism (CDM) was established to meet the requirements of the "United Nations Framework Convention on Climate Change" and the "Kyoto Protocol." According to CDM, developing countries are permitted to receive advanced technologies and funded for clean development from developed countries to implement projects on the reduction of greenhouse gas emissions (Liu et al. 2010).

Many local governments and enterprises show increasing interest in implementing CDM, and many provinces have established CDM Service Offices to deepen the cooperation between Chinese enterprises and developed countries in technologies and management. In 2010, the total numbers of CDM projects approved by the Chinese government have reached 2685 on 7 September (National Development and Reform Commission of China 2010). Compared with 135 CDM projects in 2006, notable improvement has been made. These projects in China play significant roles in helping many enterprises to obtain advanced renewable energy technologies, such as biogas technology for reducing GHG emissions.

1.3.4 Standards for Biogas Projects

The standard system for rural biogas consists of 35 criteria and 4 categories such as basic standards, product standards, technical specifications, and construction specifications. This system involved the entire processes of building biogas plants, including design, construction, operation, and facility production to guarantee the development of biogas industrialization. Table 1.2 lists the detailed standards that have been promulgated and implemented.

Table 1.2 Standards for biogas

Serial number	Standard name
GB/T 7636-87	Standard on Design of Biogas Pipelines for Rural Household
GB/T 7637-87	Operation Rules for Construction and Installation of Biogas Pipeline for Rural Household
GB/T 3606-2001	Domestic Biogas Stove
GB/T 4750-2002	Collection of Standard Design Drawings for Household Anaerobic Digesters
GB/T 4751-2002	Specification for Inspection of the Quality for Household Anaerobic Digesters
GB/T 4752-2002	Operation Rules for Construction of Household Anaerobic Digesters
GB/T 26715-2011	Biogas Valve
NY/T 90-1988	Technical Rules for Biogas Fermentation for the Rural Household
NY/T 344-1998	Household Biogas Lamp
NY/T 465-2001	Household-scaled Biogas and Integrated Farming System-specification in Design, Construction, and Use for Southern Model
NY/T 466-2001	Household-scaled Biogas and Integrated Farming System-specification in Design, Construction and Use for Northern Model
NY/T 858-2004	Biogas Pressure Meter
NY/T 859-2004	Desulfurizer of Household Engineering
NY/T 860-2004	Digester Sealing Coatings
NY/T 1220.1-2006	Technical Code for Biogas Engineering Part 1: Process Design
NY/T 1220.2-2006	Technical Code for Biogas Engineering Part 2: Design of Biogas Supply
NY/T 1220.3-2006	Technical Code for Biogas Engineering Part 3: Construction and Acceptance
NY/T 1220.4-2006	Technical Code for Biogas Engineering Part 4: Operation and Maintenance
NY/T1220.5-2006	Technical Code for Biogas Engineering Part 5: Operation and Maintenance
NY/T1221-2006	Technical Specification for Operation Maintenance and Safety of Biogas Plant in Animal and Poultry Farms
NY/T1222-2006	Criteria for Deigning of Biogas Plant in Scale Livestock and Poultry Breeding Farms
NY/T1223-2006	Biogas-powered Generating Sets
NY/T1496.1-2007	Biogas Transmission System for Rural Household Part 1: Thermoplastics Pipes
NY/T1496.2-2007	Biogas Transmission System for Rural Household Part 2: Thermoplastics Fittings
NY/T1496.3-2007	Biogas Transmission System for Rural Household Part 3: Thermoplastics Valve
NY/T1638-2008	Biogas Cooker
NY/T1639-2008	Technological Specifications for Household Anaerobic Digesters of Fiberglass Reinforced Plastics

(continued)

Table 1.2 (continued)

Serial number	Standard name
NY/T1699-2009	Technical Specifications of Biogas Digester for Domestic Sewage Treatment
NY/T1700-2009	Determination of Methane and Carbon Dioxide in Biogas-gas Chromatography
NY/T1702-2009	Technological Specifications for Domestic Sewage Treatment
NY/T1916-2010	Biogas Power Generation Technology-Criterion
NY/T1917-2010	Technical Specifications for Mobile Discharge Facilities for Digested Sludge and Slurry
NY/T667-2011	Classification of Scale for Biogas Engineering
NY/T2065-2011	Technical Code for Application of Anaerobic Dig Estate Fertilizer

1.4 Development of Biogas Industrialization

1.4.1 Introduction of Biogas Production

Anaerobic biodigester has been widely used for biogas production in China. The first household biogas digester was constructed in 1970, and then, these small rural biogas digesters become popular. For example, more than 30.5 million household biogas digesters had been constructed by the end of 2009, generating 12.4 billion m^3 biogas totally in 2009 (Liu et al. 2008; Rural Energy Industry Association of China 2009).

Biogas digesters for smallholders have experienced significant improvement in the past 40 years due to the rapid development of biogas technology. Therefore, the water-pressure digesters, which were widely used in the mid-twentieth century, were gradually replaced by standardized and efficient biogas digesters for smallholders. These new plants can not only supply energy for domestic use, but also support sustainable agricultural systems for recycling the biomass, such as the integrated pig–biogas–vegetable systems in North China (Qi et al. 2005) and livestock–biogas–fruit systems in South China (Chen 1997). The pig, cattle, and poultry manure, human sewage mixed with crop straw, fruit or vegetable residues are used as raw materials in the biodigester, which typically has a volume of 8–10 m^3 and produces between 0.1 and 0.3 m^3 gases per m^3 of digester volume per day (Qi 2003).

Gradually, the smallholder farming transformed to larger-scale farms, for example, arable farms changed to thousands of Mu and livestock farms stock also changed to thousands of pigs or cattle (Yao et al. 2002). In the eastern coastal regions or suburban areas surrounded by big cities, it is common to find medium- or large-scale live stock. For instance, in Shandong, Guangdong, and the other provinces, the location of such farms is usually in residential areas and close to water sources, where waste may cause environmental pollution (Chen et al. 2003). Meanwhile, large-scale livestock biogas requires large amount of crops and vegetables as feed and large-scale production of crops and vegetables need large amounts of organic fertilizers. Given this, medium- and large-scale biogas plants have huge potential to deal with these problems by recycling manure to the fields (Hjorth et al. 2009).

In recent years, the number of household-size biogas digesters has been gradually increasing, of which the number of medium- and large-scale biogas plants is rising rapidly. Especially during 2013–2015, driven by the investment from the central government, biogas production in rural area has developed rapidly. By the end of 2015, household biogas production had reached 419.3 million, benefiting population of 200 million. There were 110,975 biogas productions invested by central government and local government, of which 103,898 were medium-to-small-scale plants, 6737 were large-scale plants, even 34 were super-large plants, and 306 were industrial waste biogas projects. Furthermore, 458 biogas projects used crop straw as raw materials and 110,517 biogas projects used livestock and poultry manure as raw material. The total pool capacity of national rural biogas projects had reached 18,955,800 m^3 with an annual output of 2.225 billion m^3 of biogas and gas supply households reached 209.18 million.

In summary, a strategic mode of biogas production industry is depicted in Fig. 1.2, which can support the long-term energy development strategy and reduce environmental pollution.

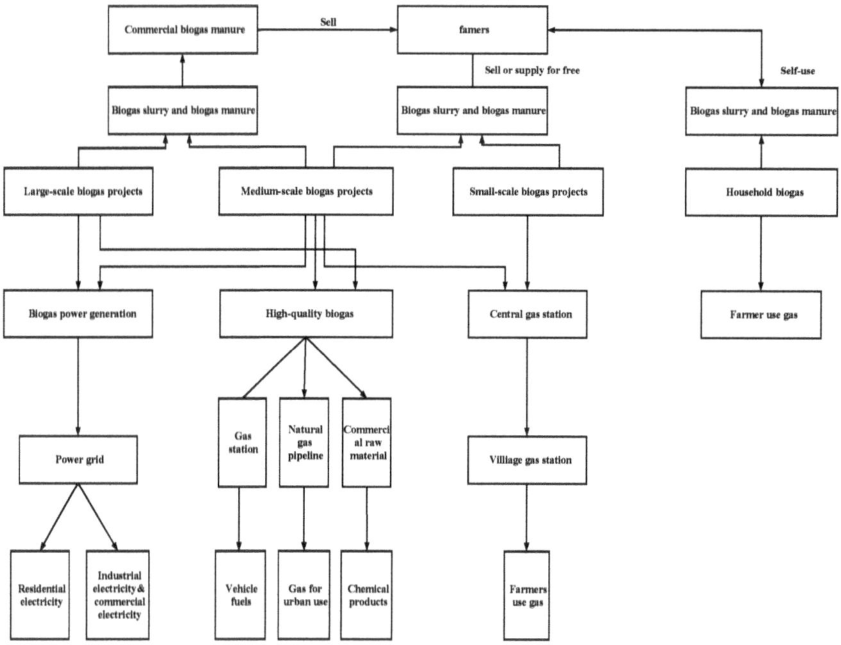

Fig. 1.2 Strategic development mode of biogas production in China

1.4.2 Biogas Plant Industry

The biogas plant industry involves a series of commercial chains, including biogas production facilities, specialized equipment, and relevant economic activities from the production and utilization of biogas and by-product treatment. It was reported that 38.51 million biogas digesters for rural households had been established until 2010 and the total output of biogas reached 13.08 billion m^3 every year. The whole biogas industry had a production ability with a total digester capacity of 8.57 million m^3 and annual output biogas of 1.05 billion m^3. It is estimated that 72,600 biogas plants had been built to digest agricultural wastes, of which 4641 were large-scale biogas plants, 22,795 were medium-scale biogas plants, and 45,259 were small-scale biogas plants. It was estimated that large-scale biogas plant, medium-scale plants, and small-scale plants can produce 613,277 and 164 million m^3, respectively.

1.4.3 Biogas Equipment Industry

The biogas equipment industry has formed mature industrial chains, involving the production of equipment, facilities, and instruments, as well as economic activities and the division of labor during the production phase. According to the statistics, more than 1300 enterprises are closely related to biogas production, which create about 23,351 jobs. Among these, there are over hundreds of companies manufacturing biogas stoves and related products. The output of these companies is quite considerable, nearly 10 million units, including stoves, pipe fittings, and desulfurization equipment (Department of Rural Economy 2009).

The production technologies of digestion equipment such as household biogas steel molds and glass fiber-reinforced plastics digesters are fully developed and fairly effective. The number of enterprises producing anaerobic digesters of various structures, such as steel structure, reinforced concrete structure, enamel-assembly structure of biogas plants, is going up quickly. Also, the performance of ancillary equipment, e.g., automatic control instrumentations and pumps, has been improved. The economic activity generated by biogas production further promotes the development of better biogas products and boosts the industry's expansion. It was reported that the total output value of this industry in 2009 with 24.7 billion RMB was over four times than that in 2005.

1.4.4 Biogas Service Industry

(1) Research and development system. Over 200,000 employees with professional and technical skills are working in biogas research field in many universities and research institutes in China.

(2) Management and development system. There are five-level agencies for managing and promoting energy in rural China (including nation, province, city, county, township). It is estimated that 12,886 administrative and technology promotion agencies are hiring over 40,000 employees in China.
(3) Quality supervision and inspection system. It consists of a series of agencies, including Quality Supervision, Inspection and Testing Center for Biogas Products and Equipment of the Ministry of Agriculture and its sub-centers. These provide many services such as quality control, effectiveness evaluations, and technical guidance; meanwhile, these agencies are supported by technicians who are responsible for products matching, quality sampling and testing, and biogas projects inspection.
(4) Training and vocational skills certification system. In 2010, 273,000 people working in biogas industry received training courses from many professional agencies, such as the Instructing Station for Rural Energy Vocational Skills Appraisal, 29 provincial-level occupational skills testing stations of industry-specific types, the Central Agricultural Broadcasting School, and the China Agricultural Society, which are led by the office of the Ministry of Agriculture. The number of technical workers who were qualified as "biogas production workers" was over 292,400 totally. All of them have been given the national vocational qualification certificate.
(5) Service system. The service system for biogas industry mainly consists of county-level service and rural service outlets. There are 756 county-level service stations and 79,177 rural service outlets built to supply biogas users with a variety of services, e.g., the biogas plant construction, operations management, and maintenance or repair.

1.5 Development Planning for Biogas Production

During the Twelfth Five-Year period, rural household biogas has contributed to energy saving in agriculture and environment protection. Due to external environment of biogas production suffering from dramatic changes, new tasks and requirements have been put forward in response to many changes in the patterns of rural production, residence, and energy-consuming. Therefore, biogas industry should deal with rising problems and tap more potential on future development.

1.5.1 Developing Motion of Biogas Production

The external environment of the construction and development on rural biogas production experienced dramatic changes over time. Even if planting industry and the urbanization were booming fast, the ratio of household biogas utilization declined generally and the demands of biogas plants were getting smaller.

Small-to-medium-sized biogas projects run not well, most of which are unable to make expected profits, and their long-term sustainable operational capacities were declining. In addition, existing biogas productions are facing various difficulties, such as the high cost of raw material security and transportation, generating a lot of liquid to dissolve, low technology of engineering and insufficient commercialization development of biogas project and end products. As a result, rural biogas system has to transfer toward the large-scale, comprehensive utilization, benefit-driven, technic innovation support for restructuring and upgrading.

1.5.2 Rural Biogas Development Policies

The central government should establish well-rounded policy supportive structure and make more efforts for biogas production because rural biogas is responsible for the disposal of rural waste, clean energy supply, environment protection, and other multiple social public welfare functions. In a long term, the financial support from the central government that is mainly used for the subsidies in the front of investment and the operation method is strong. However, the cooperation mechanism has not yet effectively established. Social capital investment in biogas project construction and operation is not sufficient with little government investment amplification effect. Besides, the supportive policies on sustainable development of rural biogas are not systematic because many policies for rural waste disposal, end products subsidies, and flows are unavailable. Since the transformation of biogas production, the demands of land, electricity, financial loan from large-scale biogas production are increasing. On the other hand, the difference between standards in various places makes it difficult for the construction, operation, and supervision of biogas projects.

References

Cao E, Rural Households use biogas up to 30.5 million and save 15.0 billion Yuan a Year. 2009. http://nc.people.com.cn/GB/61154/9957970.htmlS. (in Chinese).

Chen RJ. Livestock–biogas–fruit systems in South China. Ecol Eng. 1997;8:19–29.

Chen SQ. Analysis of large–scale management model of rural economy. J Xiangfan Univ. 2003;24(6):25–30.

Chen Y, Yang GH, Sweeney S, Feng YZ. Household biogas use in rural China: a study of opportunities and constraints. Renew Sustain Energy Rev. 2010;14:545–9.

Construction (Trial). http://www.zgny.com.cn/ifm/consultation/show.asp?ncon_id=13341S (in Chinese).

Deng L, Liu Y, Zheng D, Wang L, Pu X, Song L, Long Y. Application and development of biogas technology for the treatment of waste in China. Renew Sustain Energy Rev. 2017;70:845–51.

Feng Y, Guo Y, Yang G, Qin X, Song Z. Household biogas development in rural China: On policy support and other macro sustainable conditions. Renew Sustain Energy Rev. 2012;16(8):5617–24.

Gao YC, Kuang ZS, Pan MS, Huang XG, Chen W, Ye MQ, Huang XG, Chen W, Ye MQ, Xiao GS. Development progress and current situation analysis of the rural household biogas in China. Guangdong Agric Sci. 2006;11:22–6 (in Chinese).
Hjorth M, Nielsen AM, Nyord T, Hansen MN, Nissen P, Sommer SG. Nutrient value, odour emission and energy production of manure as influenced by anaerobic digestion and separation. Agron Sustain Deva. 2009;29(2):329–38.
Hu GQ. Status, Problems and recommendation of biogas development in rural China. Agric Eng Technol (New Energy Ind). 2008;5:15–8 (in Chinese).
Liu HG, Liu WD, Tang ZP, Fan XM. The effect analysis of regional industry structure adjustment for CO_2 emission reduction in China: on the base of inter-regional input-output method. Areal Research and Development 2010;29(3):129–35 (in Chinese).
Liu Y, Kuang YQ, Huang NS, Wu ZF, Xu LZ. Popularizing household–scale biogas digesters for rural sustainable energy development and green–house gas mitigation. Renew Energy. 2008;33 (9):2027–35.
Ministry of Agriculture and National Development and Reform Commission of China. 2004. Management Approach of National Debt project for Rural Biogas.
Ministry of Agriculture. Livestock industry development planning (2010-2020): Beijing: Science and Technology Education Department of Ministry of Agriculture, 2007 (in Chinese).
Ministry of Environmental Protection of China. 2010. http://www.zhb.gov.cn/S (in Chinese).
National Development and Reform Committee of China. 2006. Renewable Energy Prices and Cost–sharing Management Trial Procedures. (in Chinese).
National Development and Reform Commission of China. 2010. The latest approved CDM projects. http://cdm.ccchina.gov.cn/WebSite/CDM/UpFile/File2508.pdf. [Accessed in 7.9.10] (in Chinese).
National People's Congress of the People's Republic of China. 2009. Renewable Energy Law (revision) of P. R. China. http://www.npc.gov.cn/huiyi/cwh/1112/2009/12/26/content_1533216.htmS. (in Chinese).
Pachauri S, Jiang LW. The household energy transition in India and China. Energy Policy. 2008;36(11):4022–35.
Qi X, Zhang S, Wang Y, Wang R. Advantages of the integrated pig-biogas-vegetable greenhouse system in north China. Ecol Eng. 2005;24(3):175–83.
Qi ZH. Construction of circle economy and eco–town. China Popul Resour Environ. 2003;13 (5):111–4 (in Chinese).
Rural Energy Industry Association of China. China's biogas industry development report. China Rural Energy Industry Association 2009, Beijing. (in Chinese).
Sun YM, Li GX, Zhang FD, Shi CL, Sun ZJ. Status quo and developmental strategy of agricultural residues resources in China. Trans Chinese Soc Agric Eng. 2005;21(8):169–73.
Thirteen–Five planning of national rural biogas development. NDRC&MOA.201701. www.sdpc.gov.cn/zcfb/zcfbghwb/201702/W020170210515499067992.pdf, http://www.sdpc.gov.cn/zcfb/zcfbghwb/201702/W020170210515499067992.pdf.
Wang XJ, Lu XG, Yang GH, Feng YZ, Ren GX, Han XH. Development process and probable future transformations of rural biogas in China. Renew Sustain Energy Rev. 2016;55:703–12.
Yao XJ, Hao X, Guo X. Commercialized operation model and development of integrated energy-environment engineering on scaled livestock farms. Transactions of the Chinese Society of Agricultural Engineering. 2002;18(1):181–4.
Zhang B, Chen B. Sustainability accounting of a household biogas project based on emergy. Appl Energ. 2016;194:819–31.
Zhang PD, Yang YL, Tian YS, Yang XT, Zheng YH, Wang LS. Bioenergy industries development in China: dilemma and solution. Renew Sustain Energy Rev. 2009;13(9):2571–9.
Zhao ZY, Fan LL. Renewable energy policies analysis and its impact on power generation structure. Renew Energy. 2010;28(4):5–9 (in Chinese).

Chapter 2
Main Methods

Abstract In order to have a better understanding of the different aspects (economic performance, environmental impacts, sustainability) of the biogas projects, ecological–economic methods are used to formulate the integrated assessment framework. First, economic method (cost–benefit analysis) was used to assess the economic feasibility of the biogas project. Moreover, DEA method was used to make an economic efficiency assessment, based on which the optimization suggestions could be provided. In terms of the environmental impact evaluation, life-cycle assessment model was established to evaluate the energy cost and environmental impact of biogas system, based on which the key sections for energy saving and emission reduction of biogas systems could be identified. As for the sustainability analysis, considering ecological and social inputs, emergy analysis and exergy analysis were employed to quantify the environmental pressure, renewability, economic efficiency, and sustainability of biogas systems. Possible pathways to achieve sustainable and low-carbon biogas project management were also analyzed based on the scenario analysis. Finally, analytic hierarchy process (AHP) method was adopted to incorporate categories of indicators to have a comprehensive performance analysis of the biogas system.

Keywords Cost–benefit analysis · DEA · LCA · Emergy · Exergy · Analytic hierarchy process

2.1 Life-Cycle Assessment (LCA)

2.1.1 Background of LCA

Life-cycle assessment (LCA) dates from the late 1960s and early 1970s, and has experienced three development stages, i.e., the conception (1970–1990), standardization (1990–2000), and elaboration stages (2000–present) (Guinée et al. 2011).

B. Chen et al., *Biogas Systems in China*, DOI 10.1007/978-3-662-55498-2_2

Various studies were conducted in the first stage to evaluate the cumulative energy requirements for the production of products and industrial processing such as steel, pulp and paper, and petroleum refining. In the early 1970s, when oil crisis took place, extensive energy studies had been conducted for a lot of industrial systems (Fava and Page 1992). By the end of the 1980s, LCA had been widely used by private companies in European countries such as Sweden, Switzerland, and the USA (Huppes 1996; Udo de Haes 1993). However, there is a lack of a common theoretical framework for LCA studies. As a consequence, the conceptions and results of LCA were widely diverging.

With the increasing severe environmental issues threatening the economic development and human living, people's environmental consciousness was running high gradually during 1990–2000. LCA is a powerful tool in tackling these issues, by analyzing the environmental impacts from the cradle to grave or from cradle to gate. Discussion on the importance of the life cycle of products was heated during 1980s to 1990s (Guinée et al. 2011). Now, LCA is experiencing a boom in the directions of breadth, depth, and applications and has becoming a policy analysis tool worldwide. LCA experienced unprecedentedly evolution in this period (e.g., Guinée et al. 1993a, b; Ayres 1995; Finnveden 2000). Some organizations also dedicated in the standardization of LCA framework. The Society of Environmental Toxicology and Chemistry (SETAC) started becoming a leader and coordinator in improving and harmonizing LCA framework, terminology, and methodology (SETAC-Europe 1993; Fava et al. 1993). Similar efforts were also undertaken by International Organization for Standardization (ISO). International standards on LCA have been made, including ISO 14040 (LCA within environmental management), ISO 14041 (inventory analysis), ISO 14042 (impact assessment), ISO 14043 (interpretation). These standards were then rephrased to ISO 14040 and ISO 14044 in 2006. Therefore, primary LCA procedure and standardization had been accomplished in this period.

In the third stage, LCA experienced a bloom in both methodology improvement and applications. Under the basic ISO framework, diverging approaches had been proposed, which contributed to the development of LCA. For example, calculation technologies were extended from process LCA to environmental input–output-based LCA (EIO-LCA) (Hendrickson et al. 1998; Peters 2007) and hybrid LCA (Lenzen 2002; Crawford 2008). Assessment method had been extended from midpoint evaluation (CML 2002, EDIP 2003, TRACI) and endpoint methods (EPS, Ecoindicator 99) toward methods that try to combine these two approaches and model impacts at both mid- and endpoint levels (LIME, ReCiPe, IMPACT2002+). Risk assessment was also incorporated into LCA (Nishioka et al. 2006; Saouter and Feijtel 1999; Sonnemann et al. 2004). In terms of applications, LCA continued to grow in importance in national decision making in both EU (European Commission 2013) and the USA (EPA 2013).

2.1.2 Methodology of LCA

LCA is a tool to assess the potential environmental impacts and resources used throughout a product's life cycle, i.e., from raw material acquisition, via production and use phases, to waste management (ISO 2006). It is a lifetime assessment that incorporates all material and energy inputs of a specific process and assesses their impact on natural environment, human health, and resource depletion. Using LCA, the following merits can be achieved:

(1) LCA is a powerful tool in examining the environmental impacts of a product or service throughout its life cycle;
(2) LCA provides a comprehensive overview of a product or service and avoids simply shifting from one stage of the lifetime to another, from one place to another, or from one environmental issue to other ones;
(3) LCA can guide decision-making process for enterprises, sectors, and governments;
(4) LCA helps to make clear the actual life-cycle environmental impacts of a product, which is always underestimated by people.

There are three steps for a standard LCA. (i) First of all, the goal and scope of the concerned system should be determined. Within the defined research scope, all the mass and energy inputs and outputs during lifetime stages of the chosen product, covering production, utilization, and final disposal and recycling, should be compiled in the inventory; (ii) based on the inventory of inputs and outputs, environmental impacts associated with the mass and energy flows should be quantified; and (iii) interpretation of the results and finding out appropriate ways to release the environmental, economic, and other pressures (Dincer and Rosen 2007). The LCA framework proposed by ISO is demonstrated in Fig. 2.1.

Two approaches are available in quantifying the environmental impacts of a product, i.e., the process-based and environmental input–output-based LCA (EIO-LCA). The conventional process-based LCA is a bottom-up approach, which traces all environmental impacts along with the supply chain. However, there are some cutoff criteria in process-based LCA, which neglect the parts that are considered unimportant or make few contributions to the results. This may lead to an underestimation of the LCA results. According to Suh et al. (2004), the impacts of the cutoff are 20% for many impact categories. Environmental input–output-based LCA can eliminate the cutoff derived from process-based LCA, as it is based on the national account and contains national economy and imports. However, the accuracy of EIO-LCA results may be decreased due to the uncertainty generated in sectoral aggregation (Mattila et al. 2010). In addition, data used in EIO-LCA are always outdated, as national input–output tables are not published on an annual basis.

To overcome the deficiencies of these two methods, hybrid LCA, proposed by Carnegie Mellon University (CMU), was proposed as a state-of-the-art LCA (Hendrickson et al. 1998; Lave et al. 1995). The hybrid LCA represents methods

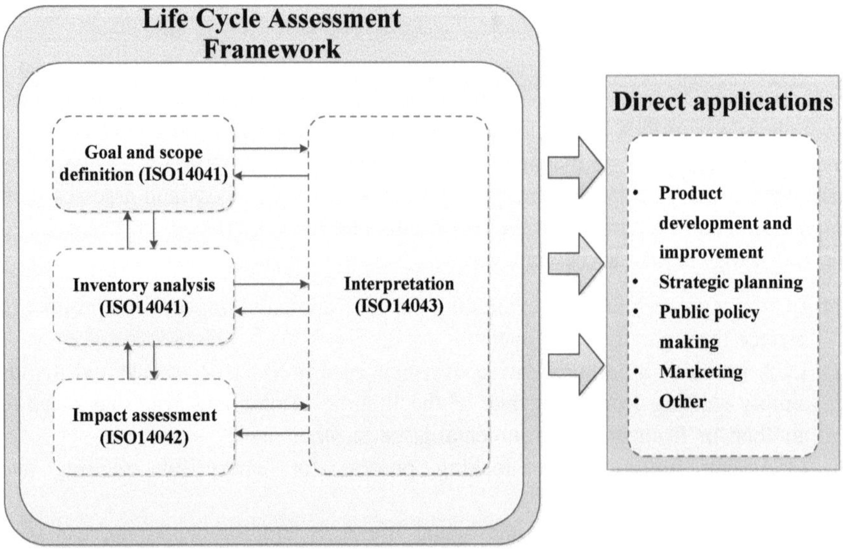

Fig. 2.1 LCA assessment framework

that combine process-based and EIO analysis to reduce uncertainty (Zhai and Williams 2010). Now, three prevailing hybrid LCA methods exist, namely additive hybrid (Bullard and Herendeen 1975), economic-balance hybrid (Williams 2004), and mixed-unit hybrid (Hawkins et al. 2007). Several real-world applications have been presented recently (e.g., Li et al. 2012; Whitaker et al. 2013), which demonstrate that hybrid LCA can avoid truncation and erroneous rankings of LCA results.

2.1.3 LCA-Based Integrated Evaluation Indices

Based on the LCA results, a multiobjective evaluation system is necessary to monitor biogas project performance from a systematic perspective. In the present work, an indicator system that is beneficial to a synthesized consideration of embodied energy, GHG emission, and economic factors in system optimization and policymaking was proposed.

Energy efficiency, GHG emission per energy output, and internal rate of return (IRR) are normally used for overall performance analysis. Distinct from these, new indicators of energy intensity and GHG emission intensity, defined as energy and GHG emission cost per unit profit, are proposed as goal functions for potential low-carbon and high-efficiency optimization of biogas systems. The calculation and implication of each indicator are shown in Eqs. (2.1)–(2.6).

Material recycling rate (MRR) is defined as the ratio of recycled materials in the dismantling phase to total material input based on the embodied energy metric, as shown in Eq. (2.1). MRR per se cannot be a proper energy indicator for sustainability issues, because it does not include the difference between thermal and mechanical energy based on the second law of thermodynamics. However, MRR may describe the material recyclability of a biogas system. The higher the MRR, the more materials recycled in the dismantling phase. Since recycled materials can be reused to substitute material input for biogas system construction, energy use and GHG emissions embodied in the construction phase would be reduced:

$$\mathrm{MRR} = E_{\mathrm{recycled}}/E_{\mathrm{in}} \tag{2.1}$$

where E_{recycled} is the embodiment of materials recycled in the dismantling stage and E_{in} is the total embodied energy input.

Energy efficiency (EE) has been frequently used as an indicator to calculate the energy budget in earlier studies. In the case of electricity generation, energy intensity entails comparison of the primary energy used in the manufacture, transportation, construction, operation, decommissioning, and other stages of a facility life cycle with the amount of electricity generated. The less energy required to produce one unit of electricity, the more efficient the biogas system. EE is calculated as:

$$\mathrm{EE} = E_{\mathrm{in}}/E_{\mathrm{out}} \tag{2.2}$$

Similarly, the GHG emission per energy output (ECD) associated with nonrenewable energy cost can be determined as:

$$\mathrm{ECD} = C_{\mathrm{in}}/E_{\mathrm{out}} \tag{2.3}$$

where C_{in} is the direct and indirect GHG emission of the production process, and E_{out} is the total energy of electricity generated by a biogas system.

New indicators of energy intensity (EI) and GHG emission intensity (CI) are defined as embodied energy and GHG emission cost per unit profit, respectively:

$$\mathrm{EI} = E_{\mathrm{in}}/(B - C) \tag{2.4}$$

$$\mathrm{CI} = C_{\mathrm{in}}/(B - C) \tag{2.5}$$

where C and B are economic costs and benefits of the biogas project, respectively.

IRR is the value of the discount rate when net present value equals zero, which can be calculated by:

$$\sum [C/(1+\mathrm{IRR})^n] = \sum [B/(1+\mathrm{IRR})^n] \tag{2.6}$$

Table 2.1 LCA-based integrated evaluation indices

Index	Equations	Implications
MRR	$E_{\text{recycled}}/E_{\text{in}}$	Material recyclability of systems
EE	$E_{\text{in}}/E_{\text{out}}$	Energy conversion efficiency
ECD	$C_{\text{in}}/E_{\text{out}}$	GHG emission per energy output
EI	$E_{\text{in}}/(B-C)$	Energy intensity
CI	$C_{\text{in}}/(B-C)$	GHG emission intensity
IRR	$\sum[C/(1+\text{IRR})^n] = \sum[B/(1+\text{IRR})^n]$	A return rate used in capital budgeting for the measurement and comparison of the profitability of economic investments

Indices proposed to evaluate the system performance of biogas system are shown in Table 2.1.

2.2 Economic Assessment

2.2.1 Cost–Benefit Analysis

Economic benefit (EB) stands for the increasing economic value for the biogas project operation. It is the incremental value comparing before-construction and after-construction of biogas project and can be expressed by:

$$\text{EB} = \sum_{i=1}^{n} \text{EB}_i \tag{2.7}$$

where EB_i represents the ith benefit ($i = 1, 2, \ldots, n$), including (1) economic benefits for substitute coal, firewood, electricity, chemical fertilizer, feed, and increased fruits; (2) environment and social benefits, such as health improvement and job creativity.

Economic cost (EC) refers to the additional economic costs during the life span of biogas system:

$$\text{EC} = \sum_{i=1}^{n} \text{EC}_i \tag{2.8}$$

where EC_i is the economic cost for the ith additional cost ($i = 1, 2, \ldots, n$), with both construction costs and management costs (fermentation, maintenance, and utilization costs) being considered.

Based on the EC and EB, four financial valuation criteria (NPV, CBR, PB, EEC) are used for economic feasibility evaluation. All the four decision criteria are included in the analysis from different aspects and can increase confidence in the viability of the investment opportunity.

Net present value (NPV) is a measurement of profit calculated by subtracting the present values of cost cash flows from the present values of benefit over a period of time. It is the sum of the net present benefits annually in the whole life span of the project, which can be given as:

$$\mathrm{NPV} = \sum_{t=0}^{n} \frac{(\mathrm{EB}_t - \mathrm{EC}_t)}{(1+r)^t} \tag{2.9}$$

where r is the discount rate and t represents the specific year within the life span of biogas project, and t_0 is the first year of biogas project (construction period). If the value of NPV > 0, the discounted benefit exceeds the discounted cost and the project would be feasible with positive benefits.

Benefit-to-cost ratio (BCR) is a description of the input–output efficiency. The formula can be explained as follows:

$$\mathrm{BCR} = \frac{\sum_{t=0}^{n} \frac{\mathrm{EB}_t}{(1+r)^t}}{\sum_{t=0}^{n} \frac{\mathrm{EC}_t}{(1+r)^t}} \tag{2.10}$$

If BCR > 1, the project is feasible with a good operation efficiency, the benefit could not make up for the cost otherwise.

Payback period (PB) is time needed to return its primary investment, which could be a reflection of the operation risk. Annual net profit is not equal every year, so the accumulated net profit in different years is calculated to find the specific year where all the primary cost is returned. The formula can be calculated by:

$$\mathrm{PB} = \mathrm{CI}/\mathrm{NP} \tag{2.11}$$

where CI represents the initial investment, and NP is the net profit every year.

Economic effectiveness coefficient (EEC) is a comprehensive indicator coordinating eco-benefit quota, eco-efficiency indicator, and operation risk index, revealing the integrated effect of the biogas project. It can be written as:

$$\mathrm{EEC} = \frac{\mathrm{NPV} * \mathrm{BCR}}{\mathrm{PB}} \tag{2.12}$$

2.2.2 Data Envelopment Analysis (DEA)

DEA's initial models called CCR (Charnes–Cooper–Rhodes) were developed by Charnes et al. (1978) to evaluate overall technical efficiency. Based on the primary models, Banker et al. (1984) introduced BCC (Banker–Charnes–Cooper) models to assess pure technical efficiency (Chen et al. 2015; Shabanpour et al. 2017). The relationship between two types of efficiency is that the overall technical efficiency equals pure technical efficiency multiplied by scale efficiency. The relationship can be expressed by:

$$\text{Overall efficiency} = Pa/Pb \tag{2.13}$$

$$\text{Pure technical efficiency} = Pc/Pb \tag{2.14}$$

$$\text{Scale efficiency} = Pa/Pc \tag{2.15}$$

In this work, the input-oriented versions were chosen for both CCR and BCC models considering that it would be much easier to improve efficiency by controlling the inputs than outputs of the biogas projects. Assume that there are z decision-making units (DMUs) converting m inputs into n outputs. For the jth DMU, $x_{ij}(i = 1, 2, \ldots, m, j = 1, 2, \ldots, k)$ inputs produce $y_{jr}(r = 1, 2, \ldots, n)$ outputs. The matrix could be expressed by Lin et al. (2015):

$$\begin{cases} x_j = (x_{1j}, x_{2j}, \ldots, x_{mj})^T, j = 1, 2, \ldots k \\ y_j = (y_{1j}, y_{2j}, \ldots, y_{nj})^T, j = 1, 2, \ldots k \\ v = (v_1, v_2, \ldots, v_m)^T \\ u = (u_1, u_2, \ldots, u_n)^T \end{cases} \tag{2.16}$$

where v represents input weights vectors and u stands for the vectors of output weights.

For the jth DMU, efficiency value can be gained through CCR-DEA model and expressed by Ma et al. (2010):

$$\begin{cases} \text{Max} \frac{u^T y_{jo}}{v^T x_{jo}} \\ s.t. \frac{u^T y_j}{v^T x_j} \leq 1 \\ v \geq 0, u \geq 0 \\ j = 1, 2 \ldots k \end{cases} \tag{2.17}$$

Model 1 can be converted into a linear programming problem through the Charnes–Cooper transformation as follows:

$$\begin{cases} \text{Max} u^T y_{jo} \\ w^T x_j - u^T y_j \geq 0 \\ w^T x_{jo} = 1 \\ w \geq 0, u \geq 0 \\ j = 1, 2 \ldots k \end{cases} \tag{2.18}$$

where $w = tv$ represents the new vectors of input weights, and $\mu = tu$ means output weights.

The model 2 can be converted into a linear programming model and non-Archimedean infinitesimal for convenient evaluation, and the equivalent CCR model could be written as model 3:

$$\begin{cases} \text{Min}(q - r(e^T s^- + e^T s^+) \\ xjhj + S^- = qxj \\ yjhj + S^- = yj \\ hj \geq 0, j = 1, 2...k \\ S^- \geq 0, S^+ \geq 0 \end{cases} \tag{2.19}$$

where q and hj are the dual variable; e^+ and e^- represent m and n dimension unit vectors, respectively; S^+ andS^- stand for the slack variables, respectively. So the judgments of the CCR dual model could be gained: If $r < 1$, the evaluated DMUs are relatively ineffective; if $r = 1$, the evaluated DMUs are relatively effective.

2.3 Emergy Analysis

2.3.1 Emergy Concept

As the biosphere is generally considered to be driven by direct solar energy and by other sources of available energy deriving from solar radiation, solar emergy, i.e., the available solar energy directly and indirectly used for a product or service, is suggested as a common measure (Campbell 1998; Odum 1996) of sustainability analysis. Emergy, developed by Odum in 1960s, is defined as the availability of energy of one kind that is used up in transformations directly and indirectly to make a product or service (Odum et al. 2000). The unit of emergy is emjoule, a unit referring to the available energy of one kind consumed in transformations. Taken sunlight, electricity, and human service as examples, all of them in different energy hierarchy can be unified on a common basis by converting them in the unit of emjoules of solar energy. The aim of emergy analysis for a specific natural or artificial system is to investigate the relationship between socioeconomic development and the natural environment, which is not limited to issues of efficient resource use and can be regarded as an attempt to fit the concerned production procedure into the multidimensional surrounding ecosystems (Yang and Chen 2014).

In emergy analysis, each form of available energy input required in the lifetime of the investigated system is converted into its solar emergy equivalent, by multiplying the available energy by an appropriate solar transformity (or unit emergy value, UEV), which is a conversion factor of available energy into emergy. Solar transformity is an indirect measure of the total support needed to generate a unit (J) of resource flow or storage. Transformity can therefore be assumed as an indicator of the position of a resource in the universal energy transformation hierarchy. The larger the transformity, the more solar energy required for the production and maintenance of the resource, product, or service of interest, and the higher its position in the energy hierarchy of the universe (Odum 1988, 1996).

As emergy analysis offers a more practical methodology to assessing the status and position of different energy carriers in the universal energy hierarchy and gives consideration to both the natural properties and economic characteristics of a system, it is widely used to evaluate public policy options and environmental impacts of renewable energy, which gives quantification of sustainable resource management questions (Chen and Chen 2012; Lapp 1991; Pereira and Ortega 2010; Yang et al. 2011). The procedure of emergy analysis includes: (1) collecting relevant ecological and socioeconomic information such as material inputs concerned with the studied biogas system; (2) determining the system boundary and main energy sources and clarifying the interrelationships among different components, based on which the emergy diagram should be drawn using emergy symbols; the components should be listed in sequence based on the their transformities, and inputs with larger transformity should be listed in the right horizontally and at the top vertically; (3) compiling emergy table. Emergy table should include input items, the quantity of energy or material inputs, solar emergy transformity, and solar emergy. (4) Some emergy-based indicators should be employed for sustainable evaluation based on the requirement of specific research.

2.3.2 *Emergy Diagram*

The emergy diagram reveals the main processes of a specific system and all material input flows to each process, feedback flows, resource degradation flows, and monetary flows. Thereby, an overview of the whole process encompassing main components and their relationships within the system could be demonstrated for a comprehensive evaluation (Dong et al. 2008). In this diagram, the inputs include renewable environmental resource (R) and nonrenewable environmental resources (N), which are the direct driving forces of a specific process from the environment; the flows of material, equipment, human labor from the economy (F) that are used for the construction, operation, and maintenance the biogas power generation system; and the system output (Y), which is the yield of the process, to which the total emergy input is assigned. The co-product output of pollutants (C) can also be embraced in the diagram. Figure 2.2 shows the commonly used symbols used in emergy diagram.

2.3.3 *Emergy Indices*

Emergy indices employed in this study derive from Brown and Ulgiati are listed below:

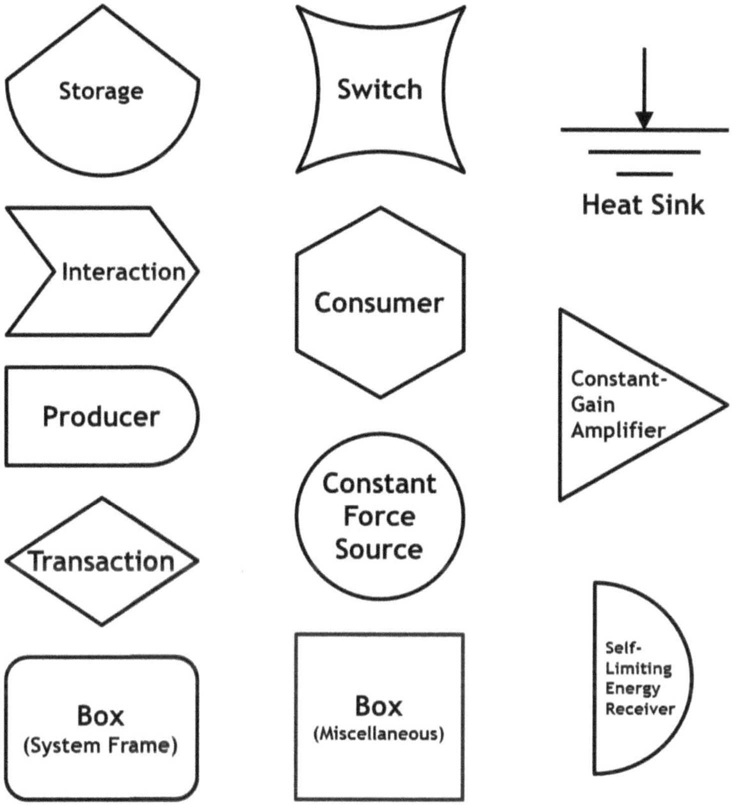

Fig. 2.2 Emergy systems language

The emergy yield ratio (EYR) is expressed as the emergy of total output Y divided by the emergy of purchased inputs outside the system of concern. The expression is EYR = Y/F. It is an indicator of the yield divided by purchased emergy input and gives a measure of the ability of the process to exploit local resources. The larger the energy yield ratio, the more output are gained, given the purchased emergy unchanged.

The environmental loading ratio (ELR) is specified as ELR = $(F + N)/R$, which is the ratio of purchased F and nonrenewable indigenous emergy N to free environmental emergy R. This indicator represents the pressure of human activities on local environment and can be used as a measurement of the environmental burden caused by human activity. If there is a relative large value of ELR, it indicates that the local ecosystem is enduring severe pressure and may induce irreversible degradation of ecosystem function.

The emergy investment ratio (EIR) is the ratio of emergy purchased from outside to the indigenous emergy inputs. It can be expressed as $F/(R + N)$. Generally, the higher the emergy investment ratio, the more money circulates and thus the higher economic development level of a system (Yang et al. 2010). It is not an independent index, but linked to the above EYR.

Emergy sustainable index (ESI) is the value of EYR divided by ELR. If a system has a high emergy yield ratio and a low environmental loading ratio, it is thereby sustainable, vice versa.

In addition to these conventional emergy indices, the system-level diversity ratio (SDR), derived from the modified Shannon information formula, is also used as an indicator to reflect the system performance and provide a quantitative assessment of the diversity.

Derived from the modified Shannon information formula, a system-level diversity ratio (SDR) was introduced by Ulgiati et al. (2011) to provide a quantitative assessment of the diversity of a system's supporting resources, defined as the ratio of actual diversity to the maximum potential diversity of a specific system. This indicator can reflect the ability to react to both biological and technological fluctuation of outputs. A lower SDR implies the output concentrates on specific kinds, which can be easily influenced by both biotic stress and market fluctuates. As a consequence, a SD closer to SD_{max} (and therefore a ratio SD/SD_{max} closer to 1) suggests higher system resilience (Table 2.2).

Table 2.2 Indicators for the energy and emergy analysis

Items	Expressions	References	Implications
Total emergy inputs (T)	$F+N+R$	Odum (1996)	Total emergy flux of the system
Percentage of renewable energy ($R\%$)	$R/(F+N+R)$	Odum (1996)	The dependence of the system on renewable emergy
Transformity	$(F+N+R)/E_{out}$	Odum (1996)	Emergy used to generate one unit of electricity
Emergy yield ratio (EYR)	Y/F	Brown and Ulgiati (1997)	Economic efficiency of a biogas system
Environmental loading ratio (ELR)	$(F+N)/R$	Brown and Ulgiati (1997)	Environmental loading exerted by the biogas system
Emergy sustainable index (ESI)	EYR/ELR	Brown and Ulgiati (1997)	Sustainability of the biogas system
System-level diversity index (SDR)	$SD = -\sum\left[\left(\frac{U_i}{U}\right)\right]\ln\left(\frac{U_i}{U}\right)$, $SDR = SD/SD_{max}$	Brown et al. (2006)	The complexity and diversity of the biogas system
E_mCO_2	$C/E_{out}/ESI$	Ju and Chen (2011)	The ratio of the real CO_2 emission released and the emergy-based sustainability indicator (ESI) per joule biodiesel

$$\mathrm{SD} = -\sum\left[\left(\frac{U_i}{U}\right)\right]\ln\left(\frac{U_i}{U}\right) \tag{2.20}$$

$$\mathrm{SDR} = \mathrm{SD}/\mathrm{SD}_{\max} \tag{2.21}$$

where U_i = emergy input of the *i*th flow = (amount of *i*th flow; J or g) × (emergy transformity of the *i*th flow; in units of sej/J or sej/g). SD is the system diversity, and $\mathrm{SD}_{\max}$ is the maximum potential diversity when the total emergy is evenly assigned to each input. As a consequence, a SD closer to $\mathrm{SD}_{\max}$ (and therefore a ratio $\mathrm{SD}/\mathrm{SD}_{\max}$ closer to 1) suggests higher system resilience.

2.3.4 Emergetic Ternary Diagrams

Graphic representation can be used for the interpretation of results in a more direct manner. Several models have been proposed to represent graphically environmental indicators (Giannetti et al. 2006). For example, Lozano (2006) offered a condensed graphical overview of the myriad of sustainability indicators. Jalal and Rogers (2002) provided a graphical representation of the state of the environment. A graphical representation of the indices obtained by emergy accounting was also reported by Brown and Ulgiati and Ulgiati and Brown (1998). Since proposed by Gibbs and Roozeboom for the analysis of mixed components, ternary diagrams have been widely used in multidisciplines (Giannetti et al. 2006). Hofstertter et al. (2000) firstly introduced ternary diagrams into ecological and environmental studies. Ternary diagram was then used to represent the interrelationships of ecosystem degradation, human health, and energy depletion. Giannetti et al. (2006) combined ternary diagram with emergy analysis and proposed an emergetic ternary diagram to shed light on sustainability management. The special data treatment and graphic representation provided by emergetic ternary diagram make it possible to compare various processes and systems, evaluate improvements, and follow the system performance over time (Chen and Chen 2012).

Emergetic ternary diagram consists of an equilateral triangle that has coordinates. The *R*, *N*, and *F* are assigned as three corners of this equilateral triangle. In the emergetic ternary diagram, the sum of proportions of *R*, *N*, and *F* is 1. Ternary combinations are represented by points within the triangle, while the relative proportions of the elements (*R*, *N*, *F*) are represented by the lengths of the perpendiculars from the given point to the side of the triangle opposite the appropriate element (Almeida et al. 2007). The resource lines and sustainability lines (Fig. 2.3) are employed to show the resource allocation in power generation systems. Detailed descriptions of emergetic ternary diagrams could be referred to Giannetti et al. (2006) and Almeida et al. (2007):

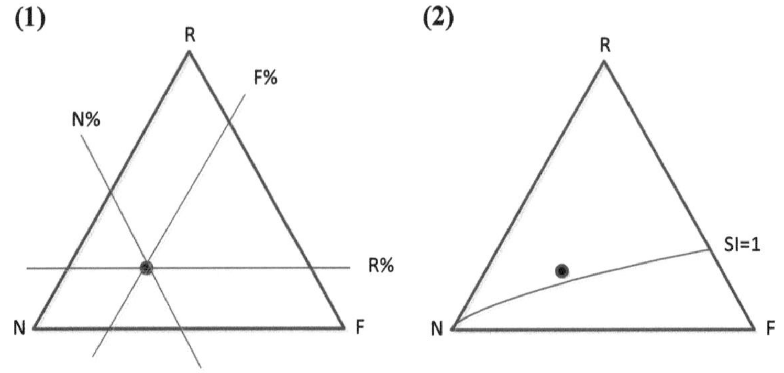

Fig. 2.3 Properties of emergetic ternary diagrams

(1) The relative proportions of the elements being given by the lengths of the perpendiculars from the given point to the side of the triangle opposite the appropriate element. These lines are parallel to the triangle sides and are very useful for comparing the use of resources by-products or processes.
(2) The graphic tool permits one to draw lines indicating constant values of the sustainability index. The sustainability lines depart from the *N* apex in the direction of the *RF* side allowing the division of the triangle into sustainability areas, which are very useful to identify and compare the sustainability of products and processes (Almeida et al. 2007).

2.4 Extended Exergy Analysis

2.4.1 Extended Exergy Analysis Framework

Extended exergy analysis is an extension of traditional exergy analysis, highlighting the primary production factors, including nonmaterial energy resource elements, labor production factors, and economic parameters. Thus, extended exergy bridges the "production value" gap between the majority of energists and economists (Chen and Chen 2009). Extended exergy (EE) intrinsically measures the amount of primary exergy homogeneously expressed in Joules that is cumulatively used over the production, operation, and disposal processes (Dai et al. 2012). The calculation of extended exergy is given by Eq. (2.20):

$$\mathrm{EE} = \mathrm{CExC} + E_{\mathrm{C}} + E_{\mathrm{W}} + E_{\mathrm{e}} \tag{2.22}$$

where EE is the total extended exergy input of a specific system, CExC is the cumulative exergy cost, E_{C} represents the exergy equivalent of the monetary flow,

E_W represents the exergy equivalent of human labor, and E_e is specified as the greenhouse gas emission abatement costs.

Extended exergy costs include three parts: (1) the standard material and energy primary resource exergy used in the lifetime of the biogas project (quantified by their respective cumulative exergy content), (2) labor flows, (3) monetary flows (two social, economic factors), and (4) greenhouse gas emission abatement costs, which are measurement of the burden of greenhouse gas emission exerted by the artificial biogas project on the atmosphere. Moreover, for cost–benefit analysis, the energy and economic outputs and greenhouse gas emission abatement benefits gained from the biogas project are also incorporated into the extended exergy accounting framework. The benefits/outputs of biogas projects include: biogas energy output, economic profits gained by the utilization of biogas and its co-products, greenhouse gas emission abatement due to the substitution of traditional biomass and fossil fuels by biogas.

Exergy consumption associated with biogas production appears not only in the process of biogas fermentation but also in the processes of delivering semi-finished products and raw materials for the biogas project. The useful energy (exergy) consumed over the life cycle is represented by cumulative exergy analysis, including nonenergetic raw material consumption (e.g., chemical energy from ore). The unit of CExC is the MJ equivalent (MJ_eq). Here, the employment of cumulative exergy analysis aims to indicate resource depletion in the biogas project.

The economic input for the biogas project is the total investment for the biogas digesters to support the construction and operation, including all construction, transportation, and operation fees. Meanwhile, economic benefits can be quantified owing to the multiple uses of biogas digestate. Economic benefits can also be attributed to the fees saved by substituting biogas for conventional energy source and substituting biogas digestate for feedstock inputs. In addition, human labor should also be taken as an economic element that swears for the functioning of the biogas project.

In terms of greenhouse gas emissions occurred in the lifetime of the biogas project, except for the greenhouse gas directly emitted onsite during biogas production process, embodied greenhouse gas emissions generated in the production and delivery of raw materials (used as inputs of the biogas project) are taken into consideration. In the evaluation of the environmental performance of the biogas project, a tradeoff between lifetime greenhouse gas emissions and emissions avoided by substituting for conventional energy should be made.

2.4.2 Extended Exergy-Based Sustainability Indexes

Taking into account the resource, economic, and greenhouse gas emission implications of extended exergy analysis, a series of indicators can be presented to reflect the conversion efficiency, renewability, carbon emission loading, economic benefits, and sustainability of biogas projects. The calculations of these indicators are shown in Eqs. (2.23)–(2.27).

(a) Resource depletion

The conversion efficiency or ε_P can be computed as the ratio of the useful output to the sum of the inputs that occurred to produce it (Wall 1977):

$$\varepsilon_P = \frac{\sum EO_j}{CExC} \tag{2.23}$$

where $\sum EO_j$ is the sum of useful resource outputs.

Renewability ($R\%$) is defined as the ratio of renewable exergy inputs to cumulative exergy inputs, i.e., the percentage of renewable energy that drives a process. In the long run, only high $R\%$ processes are sustainable.

$$R\% = E_R/CExC \tag{2.24}$$

where E_R is the renewable exergy input and CExC is the cumulative exergy input, which is the sum of renewable exergy inputs (E_R) and nonrenewable exergy inputs (E_{NR}).

(b) Greenhouse gas emission performance

Greenhouse gas emission intensity (CI) is used as the exergy equivalent to remove greenhouse gas emission (generated during the lifetime of the biogas digester) from the atmosphere divided by exergy output. CI can be a benchmark used to make tradeoffs between greenhouse gas emission and energy output.

$$CI = E_e/\sum EO_j \tag{2.25}$$

The economic return on investment (EROI) is the ratio of the economic profits gained from the biogas project to the sum of the economic investments delivered to produce it:

$$EROI = \frac{Y_C}{E_C} \tag{2.26}$$

where Y_C is the economic benefit gained by multiple utilization of biogas digestate and the substitution of conventional energy, and E_C is the exergy equivalent of the monetary inflow.

(c) Extended exergy-based sustainability indicator

Sustainable development meets the needs of economic development without compromising the environment. Taking into account both greenhouse gas emissions and economic elements in extended exergy analysis framework for biogas

Table 2.3 Evaluation indicator framework for biogas project

Destination layer	Criteria layer	Indicator layer	Explanation
	Environmental indicators	Increasing rate of organic matter	Increasing organic matter by returning biogas slurry, residue, and straw to the field
		SO_2 reduction rate	Biogas as a substitution of coal to reduce SO_2 emission
		CO_2 reduction rate	Biogas as a substitution of coal and firewood to reduce CO_2 emission
		Forest deterioration reduction rate	Biogas as a substitution of coal and firewood to slowdown forest deterioration
		Proportion of arable land taken by biogas digester	Arable land occupied by biogas digester construction
		Increase in animal excreta disposal rate	Improving residential environment quality through anaerobic fermentation
Sustainable development	Social indicators	Reduction in the rate of emigrate workers	More jobs and opportunities are provided by biogas projects
		Increase in labor productivity rate in rural areas	Surplus labor is fully used for maintenance of biogas digester
		Increase in the proportion of technical farmers	Enlarging the proportion of technical farmers through technical training on biogas technology
		Decrease in infection rate of parasitic disease	Influenced by the disposal of human and animal excreta
		Decrease in infection rate of intestine disease	Influenced by the disposal of human and animal excreta
	Economic indicators	Promotion in farmers' consuming level	Increase in income simulated the consumption
		Increasing rate of grain output per hectare	The use of organic fertilizer leads to the increase in grain output
		Increasing rate of pig breeding number	Biogas pool construction pulled the livestock breeding
		Increasing rate of fruit output per hectare	The developing mode of "pig–biogas–fruit" accelerated the development of fruit industry
		Proportion of construction cost	The proportion of pool construction costs in the whole expenditure

projects, the sustainability indicator SI can be expressed as the ratio of economic return on investment (EROI) to the greenhouse gas emission intensity (CI). The higher the index of sustainability, the lower the level of greenhouse gas emitted per unit of economic activity by the biogas project:

$$\mathrm{SI} = \mathrm{EROI}/\mathrm{CI} \tag{2.27}$$

2.5 Analytic Hierarchy Process

The multiple benefits of biogas project in rural areas cover the following: increasing the food supply by producing more and better crops, fruits, and live stocks; improving the efficiency of agricultural production; and reducing the risk of soil erosion. Meanwhile, the elimination of parasites caused by the biogas fermentation process and biogas-based cooking instead of based on dung or wood can lead to positive effects on human health. However, the construction of biogas digesters may occupy former arable land in most cases and the construction costs are unaffordable for some farmers. In consideration of both the potential impacts of biogas exerted on sustainable development and the negative influences, a tradeoff between them and thereby an indicator system should be proposed. Here, the major impacts concerning social, economic, and ecological aspects caused by the application of biogas are identified and classified in Table 2.3.

References

Almeida CMVB, Barrella FA, Giannetti BF. Emergetic ternary diagrams: five examples for application in environmental accounting for decision-making. J Clean Prod. 2007;15:63–74.

Ayres RU. Life cycle analysis: a critique. Resour Conserv Recycl. 1995;14:199–223.

Banker RD, Charnes A, Cooper WW. Some models for estimating technical and scale inefficiencies in data envelopment analysis. Manage Sci. 1984;30(9):1078–92.

Brown MT, Ulgiati S. Emergy-based indices and rations to evaluate sustainability: monitoring economies and technology toward environmentally sound innovation. Ecol Eng. 1997;9 (1–2):51–69.

Brown MT, Cohen MJ, Bardi E, Ingwersen WW. Species diversity in the Florida Everglades USA: a systems approach to calculating biodiversity. Aquat Sci. 2006;68(3):254–77.

Bullard C, Herendeen R. The energy cost of goods and services. Energy Policy. 1975;3:263–78.

Campbell DE. Emergy analysis of human carrying capacity and regional sustainability: an example using the state of Maine. Environ Monit Assess. 1998;51(1–2):531–69.

Charnes A, Cooper WW, Rhodes E. A data envelopment analysis approach to evaluation of the program follow through experiment in US public school education. Carnegie-Mellon Univ Pittsburgh Pa Management Sciences Research Group; 1978.

Chen GQ, Chen B. Extended exergy analysis of the Chinese society. Energy. 2009;34(9):1127–44.

Chen SQ, Chen B. Sustainability and future alternatives of biogas-linked agrosystem (BLAS) in China: an emergy synthesis. Renew Sustain Energy Rev. 2012;16(6):3948–59.

Chen SQ, Chen B, Fath BD. Assessing the cumulative environmental impact of hydropower construction on river systems based on energy network model. Renew Sustain Energy Rev. 2015;42:78–92.

Crawford RH. Validation of a hybrid life-cycle inventory analysis method. J Environ Manage. 2008;88:496–506.

Dai J, Fath BD, Chen B. Constructing a network of the social-economic consumption system of China using Extended Exergy Analysis. Renew Sustain Energy Rev. 2012;16(7):4796–808.

Dincer I, Rosen MA. EXERGY: energy, environment and sustainable development. Amsterdam: Elsevier; 2007.

Dong X, Ulgiati S, Yan M, Gao W. Progress, influence and perspectives of emergy theories in China, in support of environmentally sound economic development and equitable trade. Energy Policy. 2008;36(3):1019–28.

EPA (U.S. Environmental Protection Agency). Life-cycle assessment research. 2013. http://www.epa.gov/nrmrl/nrmrldatabases.htm. Accessed 20 Aug 2013.

European Commission. Joint Research Centre life cycle thinking and assessment. 2013. http://lct.jrc.ec.europa.eu/. Accessed 20 Aug 2013.

Fava JA, Page A. Application of product life cycle assessment to product stewardship and pollution prevention programs. Water Sci Technol. 1992;26(1–2):275–87.

Fava JA, Consoli FJ, Denison R, Dickson K, Mohin T, Vigon B. A conceptual framework for life cycle impact assessment, workshop report society of environmental toxicology and chemistry (SETAC); 1993.

Finnveden G. On the limitations of life cycle assessment and environmental systems analysis tools in general. Int J Life Cycle Assess. 2000;5:229–38.

Giannetti BF, Barrella FA, Almeida CMVB. A combined tool for environmental scientists and decision makers: ternary diagrams and emergy accounting. J Clean Prod. 2006;14:201–10.

Guinée JB, Udo de Haes HA, Huppes G. Quantitative life cycle assessment of products: 1: goal definition and inventory. J Clean Prod. 1993a;1(1):3–13.

Guinée JB, Heijungs R, Udo de Haes HA, Huppes G. Quantitative life cycle assessment of products: 2: classification, valuation and improvement analysis. J Clean Prod. 1993b;1(2):81–91.

Guinée JB, Heijungs R, Huppes G, Zamagin A, Masoni P, Buonamici R, Ekvall T, Rydberg T. Life cycle assessment: past, present, and future. Environ Sci Technol. 2011;45(1):90–96.

Hawkins TC, Hendrickson C, Higgins HS, Matthews HS, Suh S. A mixed-unit input-output model for environmental life cycle assessment and material flow analysis. Environ Sci Technol. 2007;41(3):1024–31.

Hendrickson C, Horvath A, Joshi S, Lave L. Economic input output models for environmental life-cycle assessment. Environ Sci Technol. 1998;32(7):184A–91A.

Hofstertter P, Braunschweig A, Mettier T, Müeller-Wenk R, Tietje O. The mixing triangle: correlation and graphical decision support for LCA-based comparisons. J Ind Ecol. 2000;3 (4):97–115.

Huppes G. LCA yesterday, today and tomorrow. Centre of environmental science: Leiden University; 1996.

ISO (International Organization for Standardization). ISO 14040 International standard. In: Environmental management—life cycle. 2006.

Jalal KF, Rogers PP. Measuring environmental performance in Asia. Ecol Ind. 2002;2:39–59.

Ju LP, Chen B. Embodied energy and emergy evaluation of a typical biodiesel production chain in China. Ecol Model. 2011;222(14):2385–92.

Lapp CW. Emergy analysis of the nuclear power system in the United States. Gaineswille: University of Florida, Department of Environmental Engineering Sciences; 1991.

Lave LB, Cobas-Flores E, Hendrickson CT, McMichael FC. Using input-output analysis to estimate economy-wide discharges. Environ Sci Technol. 1995;29(9):420–6.

Lenzen M. A guide for compiling inventories in hybrid LCA: some Australian results. J Clean Prod. 2002;10(6):545–72.

Lin W, Chen B, Xie L, Pan H. Estimating energy consumption of transport modes in China using DEA. Sustainability. 2015;7(4):4225–39.

Lozano R. A tool for a Graphical Assessment of Sustainability in Universities (GASU). J Clean Prod. 2006;14:963–72.
Li X, Feng KS, Siu YL, Hubacek K. Energy-water nexus of wind power in China: the balancing act between CO_2 emissions and water consumption. Energ Policy. 2012;45:440–8.
Ma HY, Oxley L, Gibson J, Li W. A survey of China's renewable energy economy. Renew Sustain Energy Rev. 2010;14(1):438–45.
Mattila TJ, Pakarinen S, Sokka L. Quantifying the total environmental impacts of an industrial symbiosis—a comparison of process-, hybrid and input–output life cycle assessment. Environ Sci Technol. 2010;44(11):4309–14.
Nishioka Y, Levy JI, Norris GA. Integrating air pollution, climate change, and economics in a risk-based life-cycle analysis: a case study of residential insulation. Human Ecol Risk Assess Int J. 2006;12(3):552–71.
Odum HT. Self-organization, transformity, and information. Science. 1988;242(4882):1132–36.
Odum HT. Environmental accounting: emergy and environmental decision making. New York: Wiley; 1996.
Odum HT, Brown MT, Williams SB. Handbook of emergy evaluation folio 1: introduction and global budget. Gainesville: Center for Environmental Policy, University of Florida; 2000.
Pereira CL, Ortega E. Sustainability assessment of large-scale ethanol production from sugarcane. J Clean Prod. 2010;18(1):77–82.
Peters G. Efficient algorithms for Life Cycle Assessment, input output analysis, and Monte-Carlo analysis. Int J Life Cycle Assess. 2007;12(6):373–80.
Saouter E, Feijtel TCJ. Use of life cycle analysis and environmental risk assessment in an integrated product assessment. Environmental Strategies. Nordic Workshop, Vedbaek. 1999;81–97.
SETAC. Society of Environmental Chemistry and Toxicology, Life Cycle Assessment: A Code of Practice, 1993. Available on: www.setac.org.
Shabanpour H, Yousefi S, Saen RF. Future planning for benchmarking and ranking sustainable suppliers using goal programming and robust double frontiers DEA. Transp Res Part D Transp Environ. 2017;50:129–43.
Sonnemann G, Castells F, Schuhmacher M. Integrated lifecycle and risk assessment for industrial processes. Boca Raton: Lewis Publishers; 2004.
Suh S, Lenzen M, Treloar GJ, Hondo H, Horvath A, Huppes G, Jolliet O, Klann U, Krewitt W, Moriguchi Y, Munksgaard J, Norris G. System boundary selection in life-cycle inventories using hybrid approaches. Environ Sci Technol. 2004;38(3):657–64.
Udo de Haes HA. Applications of life cycle assessment: expectations, drawbacks and perspectives. J Clean Prod. 1993;1(3–4):131–7.
Ulgiati S, Brown MT. Monitoring patterns of sustainability in natural and man-made ecosystems. Ecol Model. 1998;108:23–36.
Ulgiati S, Ascione M, Zucaro A, Campanella L. Emergy-based complexity measures in natural and social systems. Ecol Ind. 2011;11:1185–90.
Wall G. Exergy-A useful concept within resource accounting. Chalmers Tekniska Högskola Göteborgs Universitet. 1977.
Whitaker MB, Heath GA, Burkhardt JJ III, Turchi CS. Life cycle assessment of a power tower concentrating solar plant and the impacts of key design alternatives. Environ Sci Technol. 2013;47:5896–903.
Williams E. Energy intensity of computer manufacturing: hybrid assessment combining process and economic input-output methods. Environ Sci Technol. 2004;38(22):6166–74.
Yang J, Chen B. Emergy analysis of a biogas–linked agricultural system in rural China–a case study in Gongcheng Yao Autonomous County. Appl Energy. 2014;118:173–82.
Yang ZF, Jiang MM, Chen B, Zhou BJ, Chen GQ, Li SC. Solar emergy evaluation for Chinese economy. Energy Policy. 2010;38:875–86.
Yang H, Chen L, Yan ZC, Wang HL. Emergy analysis of cassava-based fuel ethanol in China. Biomass Bioenerg. 2011;35:581–9.
Zhai P, Williams ED. Dynamic hybrid life cycle assessment of energy and carbon of multicrystalline silicon photovoltaic systems. Environ Sci Technol. 2010;44(20):7950–5.

Chapter 3
Four Typical Biogas Systems in China

Abstract In this chapter, detailed descriptions of four typical systems are presented including "Six-in-One" biogas system (SIOBS), biogas-persimmon cultivation and processing system (BCPS), wastewater treatment plants (WWTPs), and the "Three-in-One" biogas project (TIOBS). The similarities and differences of how it is constructed, operated, and maintained are addressed. Besides, the unique characteristics are also depicted, containing digestate process in SIOBS, digestate reuse, and biogas utilization for persimmon cultivation in BCPS, biogas–sludge use in WWTPs, and comprehensive utilization in TIOBS.

Keywords Biogas system · Recycle use · Life cycle

3.1 "Six-in-One" Biogas System (SIOBS)

The so-called Six-in-One biogas system (SIOBS, i.e., six components compose the system including pig breeding, anaerobic digester, cropping, fruit cultivation, vegetable growing, and agricultural processing) has recently become a prevailing biogas–digestate utilization fashion in South China due to its sufficient and economic use of the fermentation by-products. The core of SIOBS is an 8 m^3 fermenting reactor, surrounded by a set of supporting devices for raw material treatment, methane utilization, and digestate recycling; Fig. 3.1 presents the overview of Gongcheng county. The field investigation of SIOBS located in Gongcheng is shown in Fig. 3.2. The SIOBS has significant advantages over the traditional agricultural practices and waste handling procedures: on one hand, it provides a clean and cheap methane fuel for electricity, heating, or machinery; on the other hand, via the fermentation of household and agricultural wastes, it is effective in processing both liquid and solid digestates into substitutions of base fertilizer, top dressing, feed additive, etc., which are otherwise high-emission chemicals.

B. Chen et al., *Biogas Systems in China*, DOI 10.1007/978-3-662-55498-2_3

Fig. 3.1 Gongcheng county located in Guangxi province, China

Fig. 3.2 "Six-in-One" biogas system (SIOBS)

3.1.1 Biogas Plant Construction Stage

The extraction and transportation of building materials for the biogas plant are covered in this stage. The application of the project has a 20-year lifespan before being discarded underground. In terms of the extraction process, cements, rebar, red brick, sands, and gravels are needed. The pipelines used in the biogas plant all adopted PVC pipes. As to the transportation of building materials, a low-speed van with a load of 12 t is assumed to be employed.

3.1.2 Biogas Plant Maintenance Stage

The biogas plant in SIOBS needs to be maintained to eliminate problems during the equipment's operation, such as water leak or steam leakage of the reactor or treatment tanks.

3.1.3 Feedstock Supply Stage

In this stage, environmental emissions are mitigated by avoiding the direct combustion of crop straws and exposure of pig manure and excrement to the atmosphere. Inoculums are also required in the process of biogas plant operation, which are usually activated sludge. As landfill is the main treatment measure of activated sludge, the reduction of relevant emissions through reusing activated sludge in SIOBS equals the magnitude of emission caused by landfill disposal.

Meanwhile, the pretreatment and transportation process of the feedstock is energy intensive. The pretreatment of raw materials should be conducted before their fermentation in the biogas plant, which needs electricity and energy consumption (Börjesson and Berglund 2007).

3.1.4 Biogas Energy Utilization Stage

The utilization of biogas not only produces efficient energy for households and other agricultural industries, but also mitigates detrimental environmental emissions by its substitution of coal and diesel. The annual output of the case SIOBS is estimated to be 450 m^3 in that the biogas plant operates 300 days a year and produces 1.5 m^3 gases per day on average. There are three utilization pathways of biogas, i.e., heating for biogas stove, illumination for biogas lamp, and alternative fuel for diesel engine, which are responsible for the substitution for coal, electricity, and diesel, respectively.

Biogas should be desulfurized before any applications. The desulfurization devices are considered in the emissions and energy consumption calculation, which contains iron and has an assumed lifetime. In the heating process, the emissions release during biogas combustion is also considered for its environmental impact. The illumination of the biogas lamp is similar to the electricity consumption of a 100 W incandescent bulb. The biogas stove and lamp cost 30 kg steels in total, which are assumed a lifetime of 10a (Jury et al. 2010).

3.1.5 Digestate Processing Stage

The digestate processing stage embraces the recycling of both liquid and solid fermentation digestates. The environmental benefits of processing digestates lie in the substitution of chemical feed additive, pesticides, top dressing, base fertilizers, and the application to seed soaking. The biogas slurry a digester produces each year amounts to 16 t, certain amount of which is utilized for substituting feed additive. Two pigs were kept by every household per year. A pig usually gains weight by 21.7% when biogas slurry is used as feed additive. Forage is needed per year. Liquid digestates are provided for the seed soaking by the digester. Compared to seed soaking treatment by using pure water, the quantity of crop production is enhanced, therefore reducing certain emissions and energy consumption from relevant cropping activities (Charles et al. 2006). Liquid digestates are used as the substitution of pesticides (for cropping or fruit/vegetable growing), whose environmental benefits are calculated from the reduction of emissions and energy consumption of conventional pesticides (Geisler et al. 2005). The rest are employed as top dressing, which work as urea, potash fertilizers, and phosphoric fertilizers. In addition, biogas residues (solid digestates) produced by fermentation are usually used as base fertilizers for the fruit/vegetable growing.

The processing activities of digestates include separation of liquid and solid digestates, granulation, and other specific treatments. The electricity and heating energy are consumed in different pathways.

3.2 Biogas-Persimmon Cultivation and Processing System (BCPS)

Agricultural system with links to anaerobic digestion prevails over many rural areas in China and has successfully evolved into different models according to diverse purposes of application, such as biogas-pig breeding-cropping and biogas-cattle breeding-vegetable growing (Wei et al. 2009; Jiang et al. 2011). One of the most successful models so far regarding the adaptation of biogas technology to local industries is the coupling of household biogas production to persimmon (*Diospyros kaki*) cultivation and processing, which was first developed and employed in the 1980s in Gongcheng Yao Autonomous County, China. Figure 3.3 provides an overview for the project. The anaerobic fermentation operates inside a small-scale digester (6 m^3, 8 m^3 or 10 m^3), in which the feedstock is the mixture of household waste and agricultural residue. Methane produced from the system is often used for direct combustion in a household family as substitutive energy for coal (biogas cooker) and electricity (biogas lamp). Besides, surplus biogas is applied to fuel the stationary machine (mainly for persimmon processing) after upgrading, serving as the substitution for diesel. Recent technology also recognizes the efficacy of CH_4/CO_2 in fruit fresh-keeping during the storage period by depressing the respiration intensity of persimmon. The fermentation by-products benefit persimmon growing by providing efficient organic fertilizers, i.e., solid digestate is utilized for early fertilization of fruit tree, and liquid digestate is used as top dressing at maturation phase. Linking biogas production to persimmon industry based on expedient installation creates a highly recycling chain of resources and therefore forms a more effective and self-reliant agro-system. The annual biogas production is up to 400–500 m^3, capable of satisfying daily energy consumption for both household daily activity and the operation of small industry. The reduced use of coal in cooking and chemical compounds in the field has brought huge environmental merits by mitigating detrimental emissions. On a yearly basis, local famers earn about $180 due to persimmon harvesting, accounting for 33% of their total incomes.

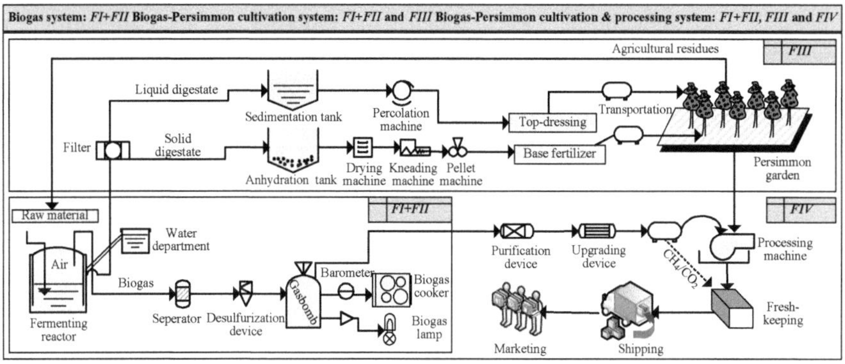

Fig. 3.3 Overview of a household biogas project coupled to persimmon cultivation and processing (*FI* Fraction I, *FII* Fraction II; *FIII* Fraction III; *FIV* Fraction IV). Reprinted from (Chen and Chen 2013), with permission from Elsevier

3.2.1 *Biogas Infrastructure (Fraction I)*

The processes of biogas construction and maintenance are considered in this fraction. The household digestion in the BCPS is conducted in an 8 m^3 cylindrical hydrostatic anaerobic digester. It is an expedient infrastructure for small-scaled biogas production that evolves from Luo Guo Rui-type biogas digesters (first set up in 1920s), consisting of two basic parts, i.e., fermentation reactor and water tank. The considered digester has a 20-year lifespan before its construction materials are disposed. Therefore, the disposal of the digester is not considered. The installation of the digester requires cement, coarse sand, fine sand, gravel, and red bricks. Also, to support the building structure, rebar is needed, in whose extraction only the steelmaking process is considered. The pipelines employed in the system are PVC pipes. During the utilization of the digester, regular maintenance is required to keep the system in normal operation and efficient production. In the maintenance process, cement, concrete, and raw materials for feedstock exchange are the basic requirement on a yearly basis. For the transportation of the above materials, a low-speed van is used.

3.2.2 *Biogas System Operation (Fraction II)*

The operation of a biogas-linked agro-system is primarily driven by anaerobic fermentation with stable feedstock. Feedstock embraces household waste (mainly pig manure and food residue) and agricultural residue (i.e., straw and organic residue collected from the orchard). In the reference system (persimmon farming system without a digester), we assume that these materials are either disposed of by direct combustion or exposed to the atmosphere. In the biogas system, the reuse of household and agricultural wastes benefits the environment by eliminating the emission from traditional disposal activities. The annual biogas produced by the household digester amounts to 450 m^3 (fluctuated from 400 to 500 m^3 during its lifetime). The combustion of methane causes slight emissions of CO_2 and N_2O, whereas its substitution for traditional energy brings significant energy and environmental benefits. There are two ways of utilizing biogas for household combustion—heating (substitution for coal) and lightening (substitution for electricity). It is assumed 1/2 of the biogas goes to heating and therefore supplies energy equivalent to 146.5 kg fire coal. One-fourth biogas is used for the purpose of lightening, and the traditional electricity is obtained from thermal power generation ($0.058 KWh^{-1}). Essential equipment to support these utilization processes includes air segregator, devulcanization device, gas bomb, biogas cooker, and biogas lamp (only the extraction of rolled steel making them is considered for simplification).

3.2.3 Digestate Reuse for Persimmon Cultivation (Fraction III)

The liquid and solid digestates from biogas digesting are utilized in the persimmon cultivation. According to the local investigation, a digester produces 6 t liquid digestate and 2 t solid digestate each year, which satisfy 70% of total fertilization in a 1200 m^2 persimmon orchard, and the other 30% are provided by commercial chemical fertilizers (while chemical fertilizers are assumed to be solely used in the reference system). In the BCPS, liquid digestate is used for top dressing of persimmon tree at maturation phase. Solid digestate produced by fermentation is employed as base fertilizers for early fertilization at tree roots. The utilization of digestate is not only good for the system by replacing the chemical fertilizers, but also is found to be effective in enhancing persimmon production. For persimmon production, the average value of Gongcheng's case is used for calculation, i.e., persimmon production increases by 200 kg by using liquid digestate, while 150 kg by using solid digestate. This leads to extra incomes of $48.78 and $36.58, respectively, when compared to the traditional persimmon production system.

There is diversity of processing techniques of biogas by-products (Rehl and Müller 2011). For simplifying the procedures, the processing of digestates only considered the infrastructure of involved machine (e.g., rolled steel) and consumption of diesel to power these machines. The basic machine of liquid digestate processing is percolation machine (10 kg rolled steel, 25 kg t^{-1} diesel for machine operation), while the processing of solid digestate is more complex, requiring drying machine, kneading machine, and pellet machine (60 kg rolled steel, 50 kg t^{-1} diesel for machine operation). In addition, additive equipment adapting the orchard to be compatible with biogas technique (e.g., special fertilizer applicator) is also accounted for (including 20 kg rolled steel). The investigation data suggest the machinery of liquid and solid digestate costs $30.50 and $39.00, while their maintenance costs $9.75 and $19.50, respectively. A household pays $2.20 for the adaptation equipment of orchard on a yearly basis. It should be noted that the machinery of digestate processing is not commonly owned by one household but shared by a couple of families. This is reflected in the ternary calculation (we assumed that five families share a set of machines). The transportation of digestate is also considered, which is similar to the calculation of transporting orchard residue (with a shipping distance of 2 km).

3.2.4 Biogas Utilization for Persimmon Processing (Fraction IV)

As mentioned, a large proportion of the produced biogas is used for daily household activities (heating and lightening). Within the system boundary of BCPS, there are two pathways to utilize biogas to a more comprehensive extent in the processing of persimmon products. The engaged processing activities before being shipped to the market include sorting of persimmons (to category them by different sizes), packaging (to pack them in labeled box), and storage (to keep fresh of the fruit during the process).

Biogas infusion has been found to be a competitive technique to keep fresh the harvested fruit over the traditional persimmon storage method. This is based on the fact that biogas can adjust the air composition of the storeroom by its richness in CH_4 and CO_2 and exiguity in O_2, and therefore depress the respiration intensity of fruits and reduce the release of ethylene. In a BCPS, only a magnitude of 10 m^3 air is needed for this purpose. The cost for this is reinforcing the storeroom with materials such as cement and concrete. And compared to the reference system, the reduction in rotten persimmon is equivalent to an increase in production of 300 kg per year, i.e., an extra income of $23.2 due to this application.

The annual production of a 1200 m^2 persimmon orchard amounts to 5180 kg (harvested once a year). A set of stationary diesel machines are responsible for conducting the processing of these products. What concerns us here is the fuel for machine operation other than the machines per se (comparative benefit). In the considered system, the surplus biogas (about 1/4 of the total volume) is utilized as the alternative fuel for diesel engines, whose offset embraces the material flows and emissions associated with the transportation and the combustion of diesel. This application reduces GHG emissions and brings direct economic income for farmers (unit price of diesel is $0.80 in China year 2011). Adaptation is needed for the processing machine to take in biogas instead of diesel. A simplified calculation is conducted due to a lack of sufficient data, in which the extraction and refinery of 4.5 kg steel and the assembling procedure are considered. In a BCPS, the location of orchard is designed to be close to local household (digester-reliant), and therefore, the distance from the orchard to market has to be increased by 10 km (the average distance from local village to the main road in Gongcheng). The ecoinvent database is used with each element modified to reflect the mass flows appropriate for the BCPS. The GHG emission coefficients and energy consumption coefficients are adapted from IPCC (2006a, b) and Berglund and Börjesson (2006). The data of persimmon production and economic benefit are derived from economic and social reports of Gongcheng Yao autonomous county and agricultural and rural economic development report of Gongcheng Yao autonomous county.

3.3 Wastewater Treatment Plants (WWTPs)

3.3.1 WWTP Construction Stage

The WWTP is one of the most popular and representative domestic plants in China, which has a working volume of 50 m^3 and capability of treating the domestic wastewater associated with 200 people. The WWTP can operate for at least 40 years. In the WWTP construction stage, the materials for building and installation and the transportation of materials were all took into considered in determining energy consumed, air emissions, and water discharges. The small-scale municipal WWTP mainly consisted of three treatment tanks: primary settler, anaerobic digester, and filter beds. The settler and digester were responsible for the wastewater digestion and purification stages where the biogas was produced. The Portland cement concrete, steel, and PVC are needed for tank construction which process consume energy and emit environmental pollutions. The filler of anaerobic tank scan usually is divided into semi-soft filler and hard filler. The denatured polyvinyl chloride wire was a type of semi-soft filler, which can be used throughout the entire life cycle of the plant without replacement. The construction materials were transported from a distance of 50 km, where a low-speed truck was used.

3.3.2 WWTP Operation and Maintenance Stage

In the WWTP operation stage, 6000 tons of domestic sewage were disposed on a yearly basis assuming that the amount of sewage was 100 L per day per capita. The COD, TN, and TP of effluent conformed to the national discharge standard for domestic wastewater (GB18918-2002). The clarification of sewage was important for reducing total nitrogen (TN), total phosphorus (TP), and chemical oxygen demand (COD), as measured between the influent and effluent. The main source of power was coal which makes the operation a high-energy phase. The operational period was 300 days, and about 6000 tons of domestic wastewaters were treated annually. The supplement and transportation of cement and concrete waste were considered in the maintenance of WWTPs. Also, sludge was cleaned out of the anaerobic tanks every year.

3.3.3 *Biogas–Sludge Use Stage*

Biogas and sludge are products (and resources, in this case) that are by-products in the wastewater treatment process. Given proper techniques and equipment, they can be reused by the plant and closing the materials cycle of resource-waste-resource. Sludge was removed from each tank every year. Annual biogas production from wastewater treatment was 630 m^3.

3.4 "Three-in-One" Biogas Project

The "Three-in-One" biogas project is the most common and representative mode of biogas utilization in southern China. With the initiation of the Rural Eco-homes project, which aims at promoting rural household biogas utilization, some provinces in southern China, such as, Anhui, Chongqing, Hubei, Hunan and Guangxi, were selected as pilots of "Three-in-One" biogas project. The socioeconomic-environmental impacts and sustainability of this "Three-in-One" biogas utilization mode should be analyzed prior to the promotion, so as to shed light on the large-scale application.

A typical "Three-in-One" biogas project links a toilet, pigpen, and plantation, the sustainability of biogas project is evaluated in this study. As shown in Fig. 3.4, human and animal manure is the feedstock of biogas fermentation in such a system. Biogas generated in biogas digester is delivered to households for use as combust. The co-products of fermentation, i.e., biogas digestate, can be used as fertilizer for grains and fruit trees. The core of this system is the construction of a biogas digestate.

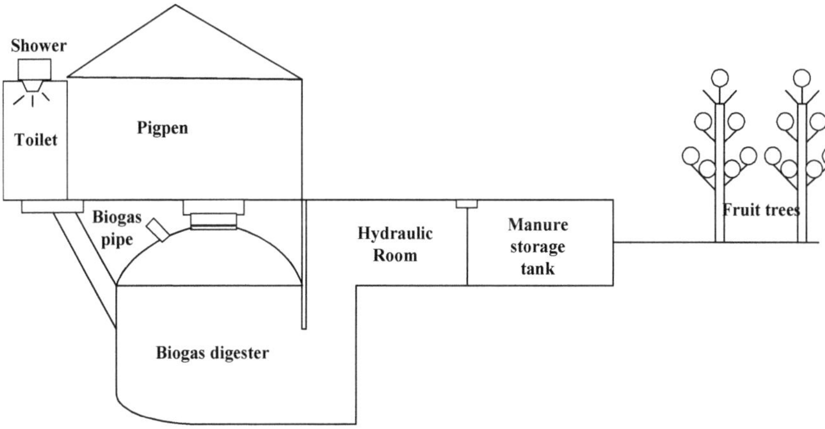

Fig. 3.4 "Three-in-One" biogas production system. Reprinted from (Yang and Chen 2014), with permission from Elsevier

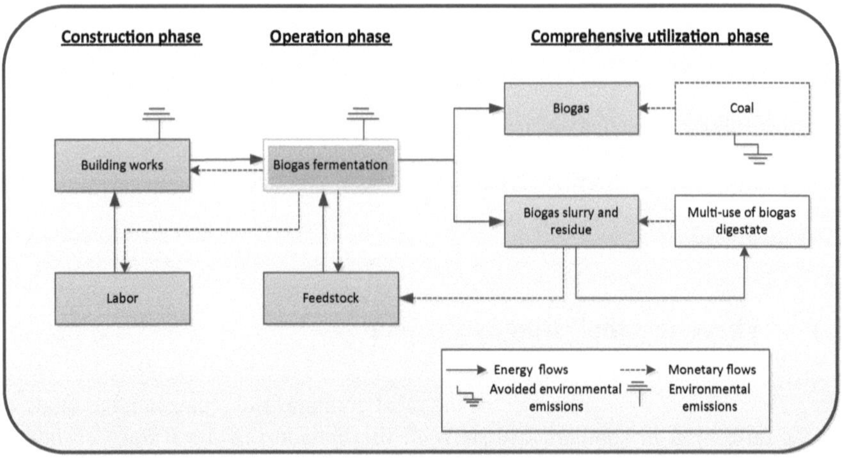

Fig. 3.5 Scheme of the evaluation of the household biogas project. Reprinted from (Yang and Chen 2014), with permission from Elsevier

Considering the convenience of management, gas generation rate, and temperature, the biogas digester should be built with the characteristics of "round shape, small capacity, and shallow." In this study, the 8 m^3 hydraulic pressure biogas digester is investigated, that digester is commonly used in southern China. The main components of such a biogas digester include a feeding pipe, fermentation room, biogas storage room, hydraulic room, and gas pipe.

Three main stages of the lifetime of the biogas production system include construction, operation, and comprehensive utilization (Fig. 3.5). These stages are further detailed in the life-cycle inventory (LCI). Energy embodied in construction materials is invested in the biogas project during the construction phase, and greenhouse gas emissions accompanied by the provision of construction materials are generated and discharged into the atmosphere. An economic input is delivered in this phase, because of the purchase of construction materials and payment for construction labor. The feedstock, such as straw and animal manure, is used to fuel the biogas production process during the operation phase. The biogas digester should be refueled once or twice per year, in order to satisfy the metabolism of methanogens, except for regular refueling every year, the fermentation feedstock should be frequently charged and discharged during routine maintenance. Gasoline is consumed for transporting fuels for the biogas digesters. The utilization of local resources is given priority to decrease transportation costs. During the comprehensive utilization phase, biogas is combusted as a substitute for conventional coal and straw; meanwhile, the biogas digestates are used for multiple purposes including breeding feed, organic fertilizer, and for soaking seeds.

References

Börjesson P, Berglund M. Environmental systems analysis of biogas systems—Part II: the environmental impact of replacing various reference systems. Biomass Bioenergy. 2007;31(5):326–44.

Börjesson P, Berglund M. Environmental systems analysis of biogas systems—Part I: fuel–cycle emissions. Biomass Bioenergy. 2006;30:469–85.

Charles R, Jolliet O, Gaillard G, Pellet D. Environmental analysis of intensity level in wheat crop production using life cycle assessment. Agric Ecosyst Environ. 2006;113:216–25.

Chen B, Chen SQ. Life cycle assessment of coupling household biogas production to agricultural industry: a case study of biogas-linked persimmon cultivation and processing system. Energy Policy. 2013;62(7):707–16.

Geisler G, Hellweg S, Hofstetter TB, Hungerbuehler K. Life-Cycle Assessment in pesticide product development: methods and case study on two plant–growth regulators from different product generations. Environ Sci Technol. 2005;39:2406–13.

IPCC. In: Eggleston HS, Buendia L, Miwa K, Ngara T, Tanabe K, editors. IPCC guidelines for national greenhouse gas inventories, prepared by the National Greenhouse Gas Inventories Programme. IGES, Japan, 2006a.

IPCC. In: Eggleston, HS, Buendia L, Miwa K, Ngara T, Tanabe K, editors. Agriculture, forestry and other land use. Prepared by the National Greenhouse Gas Inventories Programme, vol. 4. IGES, Japan, 2006b.

Jiang X, Sommer SG, Christensen KV. A review of the biogas industry in China. Energy Policy. 2011;39:6073–81.

Jury C, Benetto E, Koster D, Schmitta B, Welfring J. Life cycle assessment of biogas production by mono fermentation of energy crops and injection into the natural gas grid. Biomass Bioenergy. 2010;34(1):54–66.

Rehl T, Müller J. Life cycle assessment of biogas digestate processing technologies. Resour Conserv Recycl. 2011;56:92–104.

Wei XM, Chen B, Qu YH, Lin C, Chen GQ. Energy analysis for 'Four in One' peach production system in Beijing. Commun Nonlinear Sci Numer Simul. 2009;14:946–58.

Yang J, Chen B. Extended exergy-based sustainability accounting of a household biogas project in rural China. Energy Policy. 2014;68(2):264–72.

Chapter 4
Environment Emissions of Household Biogas Project

Abstract One of the most serious environmental problems that household biogas project would cause is the wastewater discharges and greenhouse gas emissions. Environmental discharges to the water (COD, TN, and TP) and emissions to the air (CO_2, CH_4, NH_3, SO_2, CO, NO_x, and PM_{10}) were identified. We conducted a life-cycle impact assessment (LCIA) based on the well-established mid-point methodology, Centrum voor Milieukunde Leiden (CML). The following indicators were selected: global warming potential (GWP, in g CO_2 eq), acidification potential (AP, in g SO_2 eq), eutrophication potential (EP, in g PO_4^{3-} eq), photochemical oxidation (POCP, in g C_2H_2 eq), human toxicity potential (HTP, in g 1,4-DCB eq), and abiotic depletion potential (ADP, in g Sb eq). Direct and indirect emissions are differentiated in the accounting process. The direct category includes fuel combustion in transportation, elimination of emissions from raw materials, and methane combustion as alternative household fuel, while the indirect category accounts for biogas infrastructure and equipment, digestate reuse as substitution for chemical fertilizers, machine adaptation, etc. Results show that the total greenhouse gas mitigation is 3614.157 kg CO_2-eq, to which feedstock supply stage, digestate processing stage, and biogas energy utilization stage contribute most.

Keywords Greenhouse gas emissions · Emission mitigation · Life-cycle assessment · Household biogas project

4.1 Biogas-Persimmon Cultivation and Processing System (BCPS)

4.1.1 Emissions Inventory

The annual inventory of coupling biogas production to persimmon cultivation and processing is shown in Table 4.1, covering emissions of three greenhouse gases (CO_2, CH_4, and N_2O). As previously defined, the direct and indirect categories of these aspects are clearly distinguished to provide a broader insight into the system

B. Chen et al., *Biogas Systems in China*, DOI 10.1007/978-3-662-55498-2_4

Table 4.1 Aggregated annual inventory of coupling biogas production to persimmon cultivation and processing[a]

Process	CO_2/(kg)	CH_4/(kg)	N_2O/(kg)	GHG eq-CO_2/(kg)[b]
Biogas infrastructure (Fraction I)	**−49.064**	**−0.211**	**−0.285**	**−141.726**
Direct	−10.828	−0.004	−0.046	−25.198
Indirect	−38.236	−0.207	−0.239	−116.528
Biogas system operation (Fraction II)	**932.082**	**7.213**	**3.688**	**2226.873**
Direct	929.946	5.892	3.729	2209.721
Indirect	2.136	1.320	−0.041	17.152
Digestate reuse for persimmon cultivation (Fraction III)	**−91.629**	**−0.154**	**2.213**	**560.710**
Direct	−1.895	−0.001	−0.008	−515.233
Indirect	−89.735	−0.153	2.221	1075.944
Biogas utilization for persimmon processing (Fraction IV)	**195.522**	**−0.008**	**0.817**	**448.502**
Direct	202.362	0.077	0.861	470.928
Indirect	−6.840	−0.085	−0.045	−22.426
Overall system	**986.910**	**6.840**	**6.433**	**3094.359**
Direct	1119.585	5.964	4.536	2140.217
Indirect	−132.675	0.876	1.897	954.142

[a]Negative value indicates the GHG emission, while positive value is the GHG mitigation
[b]This indicates the change of global warming potential (GWP) due to the emission or mitigation of the three greenhouse gases (CO_2, CH_4 and N_2O)

performance and different fractions compose it. In terms of the biogas infrastructure, the contributions of direct and indirect value to GHG emissions (17.78 vs. 82.22%) are revealed to be very different. It turns out the construction of biogas project caused major emissions in all three greenhouse gases during the accumulated process, while transporting the construction materials is as significant as extracting them in influencing energy consumption. The installation of a biogas digester requires $3.289 based on the current functional unit. Direct GHG mitigation (mainly due to biogas utilization) plays a dominant role in the overall profile within biogas system operation (with a contribution of 99.23%) and biogas utilization for persimmon processing (103.39%), and the same situation occurs to energy production (105.00% in Fraction II and 102.42% in Fraction IV). Basically, for these two fractions, indirect process (manufacturing essential equipments) is emission-releasing. Regarding digestate reuse for persimmon cultivation, GHG emissions are significantly reduced in an indirect way (1075.944 kg eq-CO_2, mostly attributed to the substitution for chemical fertilizers). Overall, when it comes to the accounting of coupled biogas system regarding global warming mitigation and energy production, indirect GHG emissions and energy flows play the same important role as the direct ones, that is, about half of the GHG emission mitigation and energy production are achieved in an indirect way. It should be noted that the current LCA focus on GHG emission accounting does not cover other environmental condition categories such as eutrophication, acidification, and human toxicity, but this does not mean that the biogas system has little impact on these aspects. Actually, some insights into different impact categories have been provided in the existing literatures for various types of biogas system, such as Chen et al. (2012), Börjesson and Berglund (2006), and Pöschl et al. (2012).

4.1.2 GHG Emissions Mitigation

According to the inventory result (Table 4.1), the overall mitigation of GHG emissions amounts to 3094.359 kg eq-CO_2. The building of biogas infrastructure produces emission of 141.726 kg, while the other three fractions benefit the environment by mitigating emission to different extents (2226.873 kg, 560.710 kg, and 448.502 kg in Fractions II, III, and IV, respectively). The processes contribution to GHG emissions mitigation in terms of the four fractions is scrutinized (Fig. 4.1). The construction and daily maintenance of the household biogas digester are main sources of GHG emissions in biogas infrastructure, contributing 54.12 and 28.10% of the total emission. In terms of biogas system operation, the major GHG mitigation goes to the recycling of household waste and orchard residue, and the substitution for coal follows (47.09 and 34.26%), which significantly outstrip the emission from methane combustion equipment (5.37%). The utilization of slurry (causing 50.88% mitigation in the digestate reuse stage at the expense of an increase in emission by 19.12%)turns out to be much environmentally friendly than the application of sludge (causing 16.45% mitigation at the expense of 12.73% emission). Using biogas as alternative fuel for persimmon processing contributes 92.82% GHG mitigation, while another application of biogas (utilizing biogas for fresh-keeping) results in slight GHG emissions (3.71%). By and large, eliminating the emissions from traditional waste treatments and replacing fossil fuel with

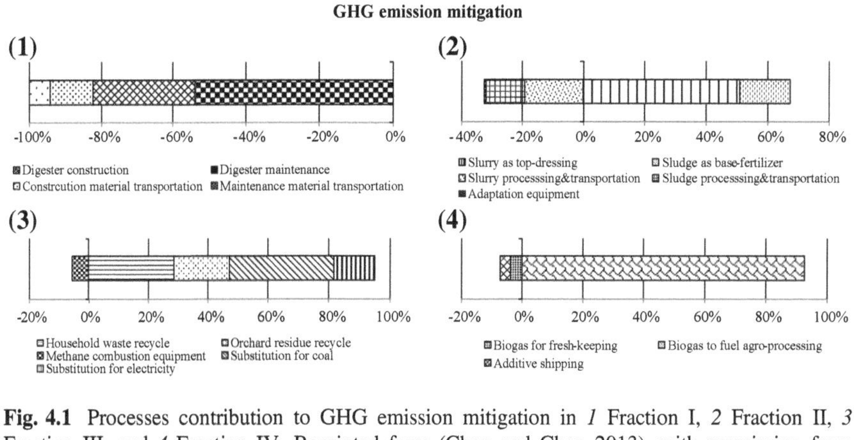

Fig. 4.1 Processes contribution to GHG emission mitigation in *1* Fraction I, *2* Fraction II, *3* Fraction III, and *4* Fraction IV. Reprinted from (Chen and Chen 2013), with permission from Elsevier

methane gives shape to the mitigation of greenhouse gases. The mitigation effect occurs whenever the household biogas digester operates (i.e., it starts to produce biogas by local wastes), which becomes even more prominent when digestate reuse for cultivation and biogas utilization for processing are applied in the system. This implicates that exploring more adaptation techniques of biogas resources is vital in offsetting the global warming potential caused by the construction and maintenance process and can be a major strategy for constructing the "low-carbon" agriculture as recommended by the Twelfth Five Year Plan in China.

4.2 Wastewater Treatment Plants (WWTPs)

4.2.1 Emissions Inventory

Table 4.2 presents the environmental emissions from a WWTP for different biogas–sludge use alternatives. The mitigation of COD, TN, and TP by the WWTP was very significant, occurring mainly during the clarification of sewage. (The elimination rates of COD, TN, and TP were 75, 25, and 62.5%, respectively). Different from COD and TP, TN was also affected in other stages. For example, the replacement of chemical pesticides in sludge disposal and processing mitigated TN discharges into the environment. Considerable GHG emissions were caused by WWTP construction, operation, and maintenance, especially during plant daily operation (>5000 g CO_2 emitted during such operations). The selection of different biogas–sludge use alternatives determined whether the process was emissions-mitigating or not. Evidently, household biogas use (HBU) and nutrient processing had the greatest GHG mitigation among all the alternatives. Also, other salient facts were apparent from the inventory results. For example, CH_4 leakage from landfill treatment of sludge was very prominent (134 g), and SO_2 was significantly reduced by HBU mainly because of the replacement of fire coal with biogas.

4.2.2 Environmental Impact Mitigation

Eight operational scenarios were developed to determine the optimal performance in terms of environmental impact for different combinations of biogas use and sludge handling. The most important criterion for selecting scenarios was to represent different types of WWTP operational patterns across the country to ensure policy relevance. The base wastewater treatment scenario (WS0) used biogas produced from anaerobic digestion solely for heating, which assumed that stored sludge was transported to a distant landfill where no household or agricultural applications were available. The second scenario (WS1) described the situation where desulfurized biogas was linked to household daily use in place of fire coal and electricity and where sludge is used for landfill because of the lack of

Table 4.2 Annual environmental emissions inventory of a WWTP

	WWTP construction	WWTP operation and maintenance		Biogas utilization			Sludge disposal and processing		
		Landfill or incineration	Nutrient processing	Heat	CHP	HBU	Landfill	Incineration	Nutrient processing
Emissions to water									
COD, g		9000.00	9000.00						
Total N, g	−0.60	300.00	300.00	0.00	−1.67	5.56		−3.79	20.94
Total P, g		150.00	150.00						
Emissions to air									
CO_2, g	−959.87	−5293.82	−5337.65	−554.40	−623.70	5304.48	−351.20	−5070.80	3326.07
CH_4, g	−0.15	−0.08	−0.08	−0.39	−0.41	7.93	−134.00		6.13
NH_3, g							−0.07		9.15
SO_2, g	−6.23	−0.57	−0.58	−0.32	−0.33	31.74	−0.30	−9.93	7.12
CO, g	−1.75	−2.06	−2.20	−1.13	−1.32	0.53		−3.80	2.86
NO_x, g	−2.06	−0.92	−0.93	−4.66	−5.73	19.07	−0.97	−13.00	40.42
PM10, g	−4.16	−1.64	−1.64	−0.22	−0.23	45.09			1.88

Note Negative values indicate environmental emissions, and positive values indicate emissions mitigation due to the activity in that stage equipment for nutrient processing. Scenario WS2 represented the combination of solid sludge incineration with biogas used for heat, and WS3 characterized the same scenario with biogas used for combined heat and power (CHP). Both scenarios included municipal WWTPs that were located close to thermal power plants or had easy access to the national grid. For scenarios of WS4 and WS5, nutrient processing was linked with CHP and HBU, respectively. These two scenarios were designed to represent systems with agricultural applications, and WS5 specifically included the most prevalent form of biogas-linked wastewater treatment used by families with biogas handling capabilities for their household biogas digesters. Since incineration and nutrient processing were associated with unique energy and environmental benefits, we considered the combination of incineration and nutrient processing (an incineration: nutrient processing ratio of 1:2) for sludge handling in WS6 and WS7. These scenarios differ only, in that WS6 assumes biogas use for CHP, whereas WS7 assumes biogas use for HBU.

Based on the emissions inventory, the environmental impact remediation by the wastewater treatment system was evaluated for different scenarios, as shown in Table 4.3. The results indicate that for global warming potential (GWP), WS5 had the best mitigation effect (almost 20 kg CO_2 eq), while WS2 and WS3 had the highest emissions. For AP, WS5 was the optimal choice for impact remediation (74.78 g SO_2 eq), followed by WS7 (55.32 g SO_2 eq), while WS2 and WS3 were associated with environmental impacts rather than mitigation. The most prominent impact remediation for HTP and ADP also occurred in WS5 (105.50 g 1,4-DCB eq for HTP and 1.03 g Sb eq for ADP). WS3, where the highest environmental impacts were encountered (-32.64 1,4-DCB eq and -0.56 g Sb eq), was the worst option of all. This indicates that the joint application of HBU and nutrient processing in WWTPs is the best option for the remediation of GWP, AP, HTP, and ADP, whereas incineration will accelerate global environmental change by constantly emitting air pollutants such as CO_2, N_2O, and SO_2. WS5 (792.69 g PO_4^{3-} eq) and WS4 (789.49 g PO_4^{3-} eq) were superior over other scenarios for mitigating EP, although the impact variations were not obvious because COD, TN, and TP were dominated by the clarification of sewage during plant operation. For POCP, the

Table 4.3 Environmental impact mitigation by WWTP under different scenarios

	GWP (in g CO_2 eq.)	AP (in g SO_2 eq.)	EP (in g PO_4^{3-} eq.)	POCP (in g C_2H_2 eq.)	HTP (in g 1,4- DCB eq.)	ADP (in g Sb eq.)
WS0	−12658.33	−11.84	780.47	−1.49	−15.99	−0.24
WS1	733.27	32.09	783.54	−0.46	52.72	0.24
WS2	−18293.35	−27.38	779.48	−1.13	−31.35	−0.54
WS3	−18695.06	−27.93	779.34	−1.18	−32.64	−0.56
WS4	5755.44	30.31	789.49	0.73	35.49	0.53
WS5	19548.75	74.78	792.69	1.82	105.50	1.03
WS6	−1970.92	11.00	786.11	0.17	12.82	0.19
WS7	11186.68	55.32	789.31	1.15	82.76	0.65

most significant remediation effect was achieved in WS5. In contrast, the most undesirable scenario was WS0 (the base system), where the combination of heat and landfill worsened the impact.

The proportional effects of different stages for impact mitigation are illustrated in Fig. 4.2. Considering the environmental impact components of the WWTP, the

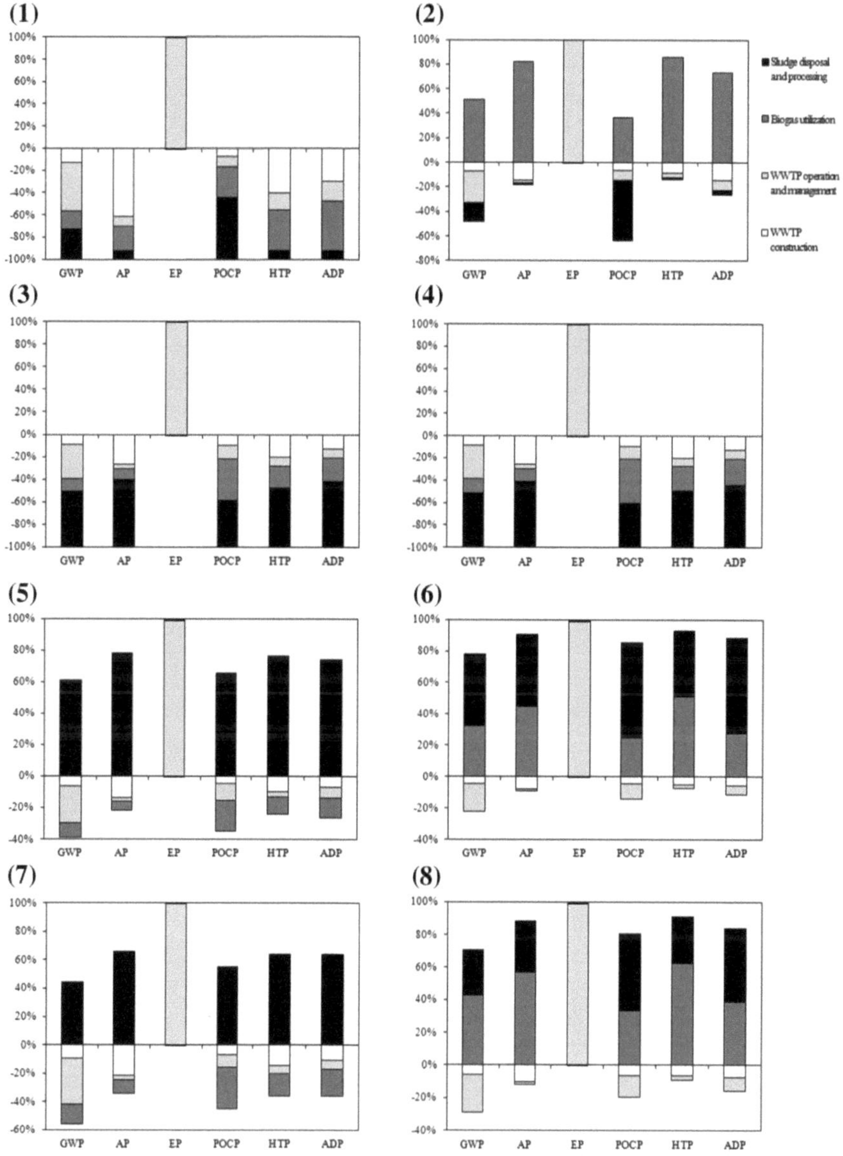

Fig. 4.2 Proportional effects of different stages in environmental impact mitigation by a WWTP for scenarios: *1* WS0, *2* WS1, *3* WS2, *4* WS3, *5* WS4, *6* WS5, *7* WS6, and *8* WS7

construction and maintenance material and plant daily operation represent the main environmental cost of the basic installation (without reuse of any resources downstream). We found that biogas and sludge use lines also had strong influence on system environmental performance (e.g., the proportional impact of heat production from biogas on HTP was 36%, while the contribution of landfilling to GWP and POCP was 27 and 56%, respectively). For WS1 (heat replaced by HBU), HBU had significant positive contributions to the mitigation of GWP, AP, HTP, and ADP (52, 81, 86, and 74%, respectively), which resulted in the WWTP becoming a source of environmental remediation for these impact categories rather than causing a negative environmental impact. For WS2 and WS3, five of the impact categories (GWP, AP, POCP, HTP, and ADP) were found to be negative. For these two scenarios, the biogas use stage accounted for over 40% of all five categories, and the sludge handling stage accounted for over 10% of these categories. Evidently, incineration was not an environmentally friendly alternative for sludge processing, and neither were heat and CHP for biogas use. A similar structure for impact mitigation was discerned in WS4 and WS6 in that CHP was used for biogas in both scenarios. However, unlike WS4 (in which all impact categories were mitigated), WS6 did not benefit the GWP because of the partial incineration of sludge. Also, WS4 had a higher remediation effect for the sludge processing stage (>60% for all impact categories) and a smaller environmental impact for biogas use (<20% for all impact categories). Thus, the entire system tended to have a more effective impact on mitigation. Regarding WS5 and WS7, all six impact categories represented environmental remediation, with both scenarios including the joint use of HBU in biogas use and nutrient processing in the agricultural industry. The difference between these two scenarios was that the WS5 sludge handling stage played a more important role for impact categories such as GWP, POCP, and ADP, while the WS7 biogas use stage was more critical in remediating environmental impacts like GWP, AP, and HTP.

4.3 'Six in One' Biogas System (SIOBS)

The goal of this study is to appraise the energy and environmental performance of the comprehensive utilization of biogas and digestate at a household scale. The considered functional unit is a SIOBS operated by an 8 m^3 cylindrical hydrostatic anaerobic digester (assumed a lifetime of 20 years) on per year basis. It should be noted that the current LCA is not based on a complete infrastructure of the SIOBS, i.e., the construction of agricultural industries is not included. Instead, in order to explore the environmental performance of the current system (with a digester) in comparison with the traditional agro-system (without a digester), the additive infrastructure and the resultant environmental benefit are calculated and demonstrated.

Figure 4.3 illustrates the system boundary of the study. In light of the principles of LCA, the inventory will include biogas plant construction, biogas plant

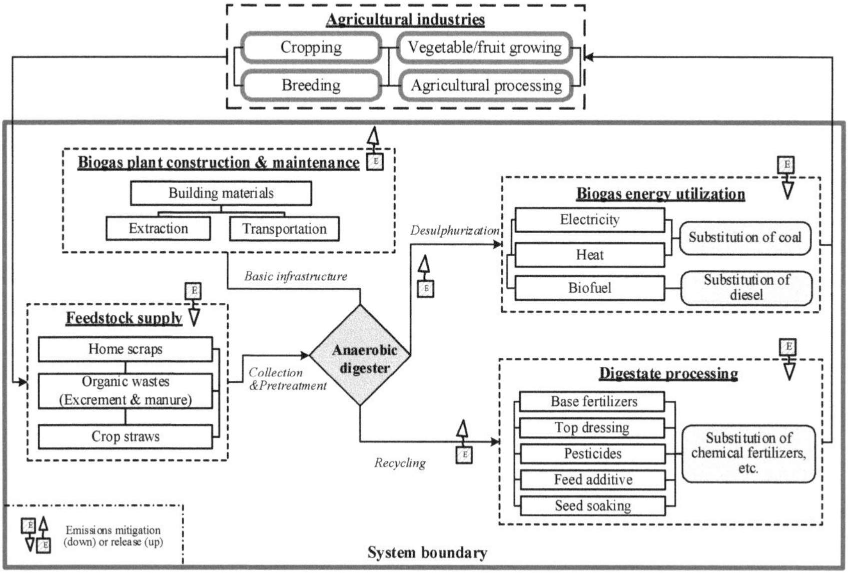

Fig. 4.3 System boundary of the study

maintenance, feedstock supply, biogas energy utilization, and digestate processing. The emissions release/mitigation is emphasized at these stages. Initial observation suggests that construction/maintenance, feedstock pretreatment, and desulphurization are emissions-released processes, while feedstock supply, biogas utilization, and digestate reuse are the major emission-mitigated processes. Within a SIOBS, the biogas project serves as an important component that links agricultural industries such as cropping, vegetation and fruit growing, pig breeding, and agricultural processing. The environmental impacts of these industries are mitigated in SIOBS due to the substitution of chemical fertilizers, pesticides, etc., with the fermentation by-products. In order to take into accounting the impacts generated by these by-products, the method for substitution estimation entails an expansion of the system boundaries. The information of the inventory is derived from multiple sources, embracing literature review, communications with local users, and inventories conducted on similar functional units and processes in the ecoinvent database (Ecoinvent 2006).

4.3.1 System Inventory

The results of the emissions inventory on a per year basis are shown in Table 4.4. The emission releases in these stages are relatively small in comparison with the magnitudes of emissions mitigation. In terms of the three major greenhouse gases (i.e., CO_2, CH_4, and CO), feedstock supply stage contributes most to the

Table 4.4 Life-cycle emissions inventory of a SIOBS (on per year basis)

	Biogas plant construction	Biogas plant maintenance	Feedstock supply	Biogas energy utilization	Digestate processing	Whole system
CO_2/(kg)	−83.374	−56.191	1222.536	1001.872	1041.005	3125.848
CH_4/(kg)	−0.009	−0.001	13.140	1.142	3.981	18.254
CO/(kg)	−0.150	−0.045	73.600	1.188	1.159	75.752
SO_2/(kg)	−0.199	−0.016	0.329	4.518	8.138	12.770
NO_X/(kg)	−0.162	−0.077	2.029	7.159	4.505	13.454
NH_3/(kg)	0.000	0.000	6.222	0.000	1.174	7.396
TN/(kg)	0.000	0.000	1.008	0.000	0.000	1.008
TP/(kg)	0.000	0.000	0.854	0.000	0.000	0.854
COD/(kg)	−0.003	−0.002	15.500	−0.001	0.000	15.495
VOC/(kg)	−0.003	−0.001	12.560	−0.025	0.343	12.874
PM_{10}/(kg)	−0.284	−0.102	4.616	6.711	3.240	14.182

Negative values indicate emissions release or energy consumption, while positive ones indicate emissions mitigation or energy production. "0.000" signifies no or negligible emission/energy

emissions mitigation, followed by stages of digestate processing and biogas energy utilization. The total mitigation of CO_2 alone in the SIOBS amounts to 3125.848 kg. Feedstock supply is also the biggest contributor to the mitigation of NH_3 and COD, while digestate processing benefits the system most by reducing the emissions of NO_X and PM_{10}.

4.3.2 Overall Environmental Evaluation

Table 4.5 displays the improvement of environmental performance by SIOBS. The contribution of different processes in SIOBS to the environmental condition improvement is shown in Fig. 4.4. The improvement of environmental condition at each stage of the SIOBS is estimated through various impact categories, including global warming (GWP) (indicated by CO_2-eq), acidification (AP) (SO_2-eq), eutrophication (EP) (PO_4-eq), photochemical oxidation (POCP) (C_2H_4-eq), and human toxicity (HTP) (1,4-DCB eq). Evidently, the improvement of environmental conditions from the feedstock supply, biogas utilization, and digestate processing is significant, attributed to the indirect benefits of these processes (Börjesson and Berglund 2007). In comparison, the impact of construction and maintenance of biogas project on the environment is relatively small. The total reduction of GHGs in the SIOBS is 3614.157 kg CO_2-eq. In particular, the mitigation of global warming benefits from the three stages during the system's operation, i.e., 1012.385 kg in the feedstock supply stage, 1012.385 kg in the biogas utilization stage, and 1096.258 kg in the digestate processing stage, which contribute 45.53, 28.01, and 30.33% to the total effect, respectively. Feedstock supply is the largest contributor to the mitigation of almost all categories (also found in Liu et al. 2008 and

Table 4.5 Improvement of environmental performance by SIOBS

Stage	GWP CO_2-eq/ (kg)	AP SO_2-eq/ (kg)	EP PO_4-eq/ (kg)	POCP C_2H_4-eq/ (kg)	HTP 1,4-DCB eq/ (kg)
Biogas plant construction	−83.870	−0.313	−0.021	−0.046	−0.090
Biogas plant maintenance	−56.292	−0.069	−0.010	−0.014	−0.015
Feedstock supply	1645.676	13.509	7.028	29.708	0.564
Biogas energy utilization	1012.385	9.201	0.900	0.065	1.924
Digestate processing	1096.258	7.157	0.905	0.796	6.142
Whole system	3614.157	29.485	8.802	30.508	8.525

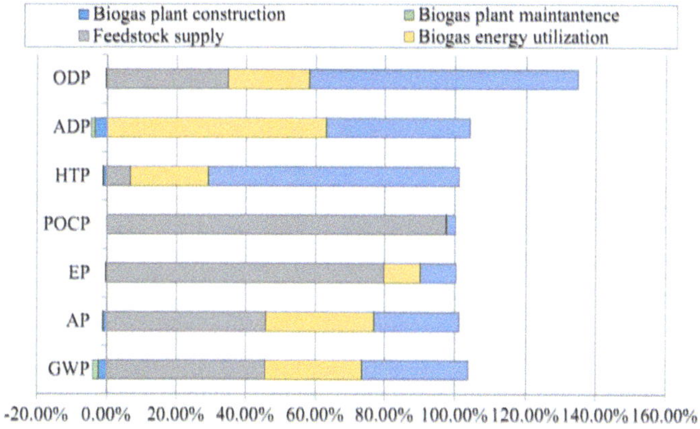

Fig. 4.4 Contribution of different stages to the environmental condition improvement

Pöschl et al. 2010), e.g., 45.82% to AP, 79.85% to EP, and 97.38% to POCP, with an exception in HTP, where the major contribution comes from the utilization terminals of the system (72.05% from biogas utilization and digestate processing from 22.57%). It is important to note that the figures are not the absolute emissions that specify the environmental performance; rather, it designates the improvement of a SIOBS over the conventional agricultural activities or waste treatments by adding a biogas project for a household.

4.3.3 *Sensitivity Analysis*

In order to identify the effect of some possible changes on the GHGs mitigation, we carried out a sensitivity analysis. Taking the utilization of biogas for direct

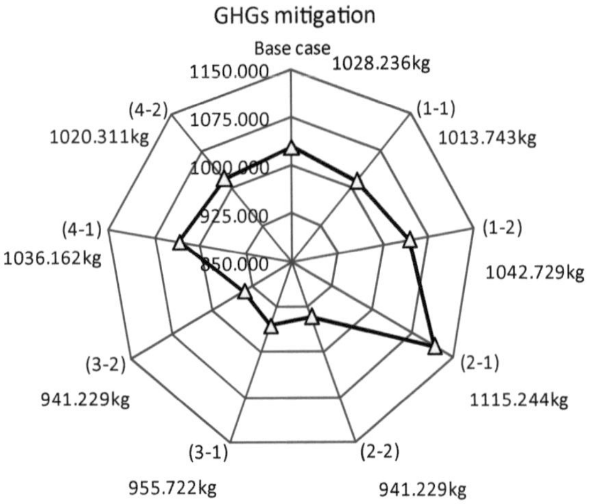

Fig. 4.5 Sensitivity analysis of GHGs mitigation

combustion (substitution of coal), powering the biogas lamp (substitution of electricity) and fueling a machine (substitution of diesel) in a ratio of 2:1:1 as the base case, six valuations were considered in this sensitivity analysis: (1-1) decreasing direct combustion by 1/6 (i.e., a ratio of 1:1:1 among the three utilization pathways); (1-2) increasing direct combustion by 1/6 (a ratio of 4:1:1); (2-1) decreasing the powering of biogas lamp by 1/6 (a ratio of 7:1:4); (2-2) increasing the powering of biogas lamp by 1/6 (a ratio of 5:5:2); (3-1) decreasing the fueling of a machine by 1/6 (a ratio of 7:4:1); (3-2) increasing the fueling of a machine by 1/6 (a ratio of 5:2:5); (4-1) reducing the cost of biogas equipment by 1/2; (4-2) adding the cost of biogas equipment by 1/2. The results of the sensitivity analysis of GHGs mitigation are shown in Fig. 4.5.

It turns out that the influences of these activities on the GHGs mitigation are disparate in the biogas utilization stage. Alterations on substitution of electricity and substitution of coal will significantly affect the mitigation of global warming. Specifically, if the biogas utilized in powering the biogas lamp is added by 1/6, GHGs mitigation will reach 1115.244 kg CO_2-eq. In comparison, the biogas equipment and substitution of diesel are insignificant factors in this context.

4.3.4 Digestate Reuse

The reuse of fermentation digestate is another major application of SIOBS that links different agricultural industries in a more efficient way and replaces the conventional disposal of household wastes (Rehl and Müller 2011). To appraise the improvement of environmental conditions by utilizing digestates in multiple pathways, the contributions of various activities to the impact assessment are

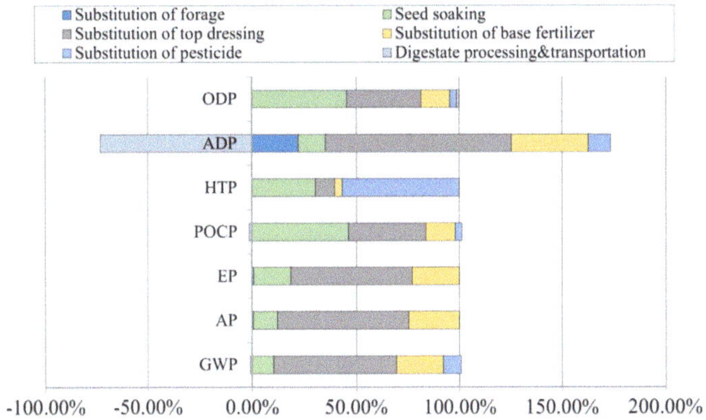

Fig. 4.6 Processes contribution to environmental condition improvement within digestate processing stage

presented (Fig. 4.6). The results show that substitution of top dressing with liquid digestate contributes over half of the total mitigation of GWP, AP, and EP (59.32, 63.33, and 58.72%, respectively). In mitigating POCP, using digestate for seed soaking is revealed to be the most important (46.16%). As to the amending of HTP, substitution of pesticide is a major contributor within the entire stage (56.98%). Evidently, in the current utilization mode, the replacement of the chemical fertilizers plays an important role in improving environmental conditions (also supported by Ahlgren et al. 2010; Pöschl et al. 2012).

GHGs reduction within digestate processing stage is shown in Table 4.6. The decrement of GHGs is 1096.258 kg, constituting 30.20% of the total mitigation in the whole system. Substitution of top dressing contributes most to the GHGs reduction in this stage (59.32%), followed by the substitution of base fertilizer (22.71%) and seed soaking (10.22%). The carbon trading prices of these GHGs reductions are also estimated, summed up as $14.20. On the other side, the utilization of digestate is less likely to augment the overall energy production.

Table 4.6 GHGs mitigation within digestate processing stage

Item	CO_2−eq/(kg)	% (%)
Substitution of top dressing	650.321	59.32(17.13)
Substitution of base fertilizer	248.966	22.71(6.56)
Substitution of pesticide	95.000	8.67(2.50)
Substitution of forage	–	–
Biogas slurry for seed soaking	112.014	10.22(2.95)
Digestate handling and transportation	−10.043	−0.92(6.05)
Total	1096.258	100.00(30.20)

4.3.5 Environmental Benefit Under Different Scenarios

With the joint operation of all five digestate processing pathways as the base case, four scenarios are further developed to determine the optimal environmental performance for SIOBS: Scenario 1: substitution of base fertilizer and top dressing; Scenario 2: substitution of base fertilizer and top dressing and forage; Scenario 3: substitution of base fertilizer and top dressing and application of biogas slurry to seed soaking; Scenario 4: substitution of base fertilizer, top dressing and fertilizer, and application to seed soaking. The environmental performance is inspected through its influences on GWP, HTP, POCP, AP, and EP.

Figure 4.7 shows the environmental performance of the SOIBS under different scenarios compared with the base case. With consideration of the benefit of digestate reuse and the cost of processing them, it turns out the judgment on the environmental conditions' pathways could be complex in terms of diverse utilization (Börjesson and Berglund 2007; Pöschl et al. 2010). That is, there might be no such an option as is better than others in every respect. Scenario 3 (combining the substitution of top dressing and base fertilizer and the application to seed soaking.) and Scenario 2 (combining the substitution of top dressing and base fertilizer and substitution of forage) can be candidates for system optimization due to the notable augment of emissions mitigation in most impact categories. In particular, GHGs reduction can be increased by 6.56 and 4.77% in Scenario 3 and Scenario 2 in comparison with the base case. Scenario 4 is most similar to the base case as constructed, but it appears to be less effective for improving the system's overall performance despite the fact that it is slightly more beneficial in remedying AP (1.51%) and EP (1.40%). Monotonic treatment of digestate for top dressing and base fertilizer (i.e., Scenario 1) is revealed unfriendly for enhancing the system's performance (also found in Jury et al. 2010; Pöschl et al. 2010). It is suggested that joint processing of digestate by multiple household units be promoted in future to lower the resultant emissions.

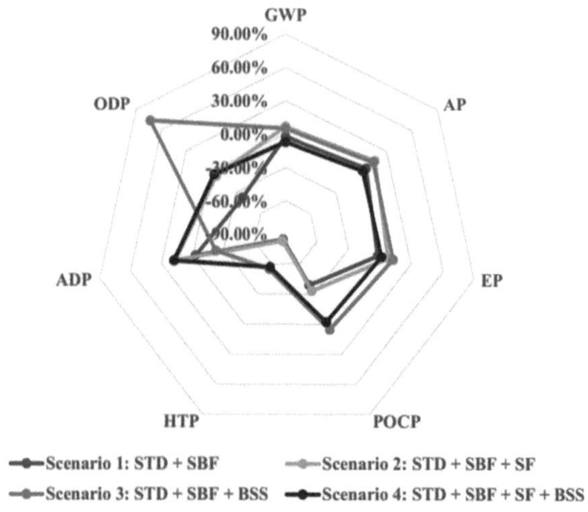

Fig. 4.7 Environmental condition improvements under different digestate reuse scenarios

STD: Substitution of top dressing; SBF: Substitution of base fertilizer; SP: Substitution of pesticide; SF: Substitution of forage; BSS: Biogas slurry for seed soaking. The proportions indicate the deviation rates from the values in base case (negative proportion denotes a decrement, and positive proportion denotes an increment).

References

Ahlgren S, Bernesson S, Nordberg Å, Hansson P. Nitrogen fertiliser production based on biogas—energy input, environmental impact and land use. Biores Technol. 2010;101:7181–4.

Börjesson P, Berglund M. Environmental systems analysis of biogas systems—Part I: Fuel-cycle emissions. Biomass Bioenergy. 2006;30:469–85.

Börjesson P, Berglund M. Environmental systems analysis of biogassystems—Part II: The environmental impact of replacing various reference systems. Biomass Bioenergy. 2007;31:326–44.

Chen B, Chen SQ. Life cycle assessment of coupling household biogas production to agricultural industry: a case study of biogas-linked persimmon cultivation and processing system. Energy Policy. 2013;62(7):707–16.

Chen SQ, Chen B, Song D. Life—cycle energy production and emissions mitigation by comprehensive biogas—digestate utilization. Biores Technol. 2012;114:357–64.

Ecoivent. Swiss center for life cycle inventories. Switzerland 2006.

Jury C, Benetto E, Koster D, Schmitta B, Welfring J. Life cycle assessment of biogas production by mono fermentation of energy crops and injection into the natural gas grid. Biomass Bioenerg. 2010;34(1):54–66.

Liu Y, Kuang Y, Huang N, Wu Z, Xu L. Popularizing household—scale biogas digesters for rural sustainableenergy development and greenhouse gas mitigation. Renewable Energy. 2008;33:2027–35.

Pöschl M, Ward S, Owende P. Evaluation of energy efficiency of various biogas production and utilization pathways. Appl Energy. 2010;87:3305–21.

Pöschl M, Ward S, Owende P. Environmental impacts of biogas deployment—Part II: life cycle assessment of multiple production and utilization pathways. J Clean Prod. 2012;24(3): 184–201.

Rehl T, Müller J. Life cycle assessment of biogas digestate processing technologies. Resource Conservation and Recycling. 2011;56:92–104.

Chapter 5
Energy Evaluation of Household Biogas Project

Abstract Biogas plant construction has been boosted in rural China not only due to the immediate merit from biogas production but also due to the succeeding benefit from by-product utilization in agro-industry, both of which are significant strategies to address energy shortage and global warming issues. Combined heat and power and household utilization were prior options in achieving net energy production, provided that power conversion efficiency and biogas production remained at satisfactory levels. A life-cycle energy evaluation of a biogas–digestate utilization system is performed. Besides, energy inputs and outputs within the system boundary are calculated to address the possibility of net energy production by wastewater treatment combined with biogas–sludge use technologies based on input–output analysis. Taking all the life-cycle stages into account, the net energy gain of the entire system is 10,458.024 MJ. Different bioresource use alternatives had a bias toward the efficiency of energy production.

Keywords Net energy production · Energy flow · Biogas utilization system · Input–output analysis

5.1 Life-Cycle Energy Evaluation

Regarding the energy performance, there are two categories in the life-cycle evaluation concerning direct energy flow and indirect energy flow. The energy consumption includes direct energy input (e.g., the heating value of diesel in transportation and machine operation) and indirect energy input (e.g., the energy required for biogas plant construction, equipment of biogas utilization, digestate processing, and machine adaptation in persimmon orchard). With regard to the energy production, direct energy output (energy produced by methane as a substitution for coal and electricity) and indirect energy output (e.g., the energy savings by replacing chemical fertilizer with digestate) are considered (also described in Chen et al. 2012). Similarly, the extending effect is defined by comparing the

B. Chen et al., *Biogas Systems in China*, DOI 10.1007/978-3-662-55498-2_5

energy production of the farming system with additive use of a certain biogas-related technique with the standard biogas operation.

The total energy input and output of biogas system is calculated in Eqs. 5.1 and 5.2, respectively. Two indices, net energy production (NEP) (Eq. 5.3) and net energy ratio (NER) (Eq. 5.4), are used for further recognition of overall energy performance. They reveal the systemic energy efficiency in two different forms.

$$E_{ti} = E_i + E'_i \tag{5.1}$$

$$E_{to} = E_o + E'_o \tag{5.2}$$

$$NEP = E_{to} - E_{ti} \tag{5.3}$$

$$NER = E_{to}/E_{ti} \tag{5.4}$$

in which E_{ti} denotes the total energy input, as a combination of direct energy input (E_i) and indirect energy input (E'_i), while E_{to} represents the total energy output, as a combination of direct energy output (E_o) and indirect energy output (E'_o).

5.2 Case Study of BCPS

The annual inventory of coupling biogas production to persimmon cultivation and processing is shown in Table 5.1. The direct and indirect categories of these aspects are clearly distinguished to provide a broader insight into the system performance,

Table 5.1 Energy inventory of coupling biogas production to persimmon cultivation and processing

Process	Energy flow/(MJ)
Biogas infrastructure (Fraction I)	**−312.013**
Direct	−166.255
Indirect	−145.758
Biogas system operation (Fraction II)	**2302.380**
Direct	2380.507
Indirect	−78.127
Digestate reuse for persimmon cultivation (Fraction III)	**−232.558**
Direct	−1710.069
Indirect	1477.511
Biogas utilization for persimmon processing (Fraction IV)	**1495.438**
Direct	1531.676
Indirect	−36.237
Overall system	**3253.247**
Direct	2035.859
Indirect	1217.388

and different fractions compose it. In terms of the biogas infrastructure, the contributions of direct and indirect value to energy flow (53.28 vs. 46.72%) are revealed to be very different. It turns out that transporting the construction materials is influencing energy consumption. Direct energy production (mainly due to biogas utilization) plays a dominant role in the overall profile within biogas system operation (with a contribution of 105.00%) and biogas utilization for persimmon processing (102.42%). Basically, for these two fractions, indirect process (manufacturing essential equipment) is energy-consuming. The processing of digestate before fertilization consumes energy (−1710.069 MJ) slightly surpasses the indirect energy production caused by the substitution (1477.511 MJ). Overall, when it comes to the accounting of coupled biogas system regarding energy production, indirect energy flows play the same important role as the direct ones; i.e., about half of the energy production is achieved in an indirect way.

The overall NEP of the system amounts to 3253.247 MJ. Biogas infrastructure and digestate reuse for persimmon cultivation are energy-intensive (NEP (Fraction I) = −312.013 MJ; NEP (Fraction III) = 232.558 MJ), while the two fractions involving biogas utilization produce energy to different extents (NEP (Fractions II) = 2302.380 MJ; (Fractions IV) = 1495.438 MJ). A high energy efficiency is observed in the BCPS (the NERs of biogas production (incorporating biogas construction and operation), digestate reuse, and biogas utilization for persimmon farming are 8.56, 2.06, and 12.92, respectively). As shown in Fig. 5.1, the energy consumed by biogas infrastructure is characterized by 73.96% in the construction process and 26.04% in the maintenance activities. Substitution for coal and electricity produces considerable energy (97.07%) at very low expense. Similar with the environmental emission aspect, the utilization of slurry (NER = 1.09) is much energy-saving than the application of sludge (NER = 0.55). Utilizing biogas to fuel agro-processing is a key process that benefits the energy performance in utilizing biogas for persimmon processing. Net energy production can be attained during the biogas plant operation, whose magnitude is increased when biogas utilization for

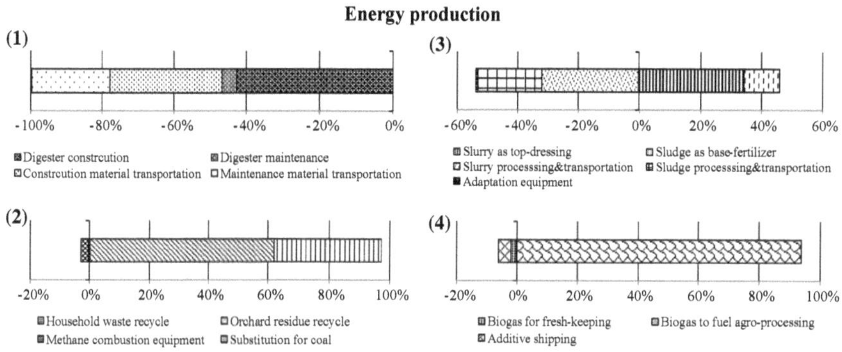

Fig. 5.1 Processes' contribution to energy production in *1* Fraction I, *2* Fraction II, *3* Fraction III, and *4* Fraction IV. Reprinted from (Chen and Chen 2013a), with permission from Elsevier

processing is employed in the production chain. However, the digestate reuse in cultivation slightly reduces the energy production of the system due to the intensive treatment of the by-products before any utilization. Obviously, this is the compromising part one has to make to extend the application of biogas resources to agro-systems (some discussion on this can also be found in Chen et al. (2012), Chen and Chen (2013a, b), Pöschl et al. (2010).

5.3 "Six in One" Biogas System (SIOBS)

5.3.1 Overall Energy Evaluation

Table 5.2 displays the improvement of energy performance by SIOBS. The energy performance of the system is indexed by the NEP. The construction and maintenance, feedstock supply, and digestate processing are overall energy-consuming, whereas biogas utilization is the energy-producing stage. With all the life-cycle stages taken into account, the NEP of the entire system is 10,458.024 MJ. Biogas utilization is the dominant process that affects the final energy supply, contributing 110.28% to the SIOBS's NEP. On the other hand, the construction process decreases the systemic NEP by 5.94%, while the recycling of liquid and solid digestate reduces it by 2.16%. It designates the improvement of a SIOBS over the conventional agricultural activities or waste treatments by adding a biogas project for a household.

5.3.2 Sensitivity Analysis

In order to identify the effect of some possible changes on NEP, a sensitivity analysis was carried out. Taking the utilization of biogas for direct combustion (substitution of coal), powering the biogas lamp (substitution of electricity), and fueling a machine (substitution of diesel) in a ratio of 2:1:1 as the base case, eight valuations were considered in this sensitivity analysis: (1-1) decreasing direct combustion by 1/6 (i.e., a ratio of 1:1:1 among the three utilization pathways); (1-2) increasing direct combustion by 1/6 (a ratio of 4:1:1); (2-1) decreasing the powering

Table 5.2 Improvement of energy performance by SIOBS

Stage	NEP/(MJ)
Biogas plant construction	−621.706
Biogas plant maintenance	−196.014
Feedstock supply	−30.492
Biogas energy utilization	11,532.604
Digestate processing	−226.367
Whole system	10,458.024

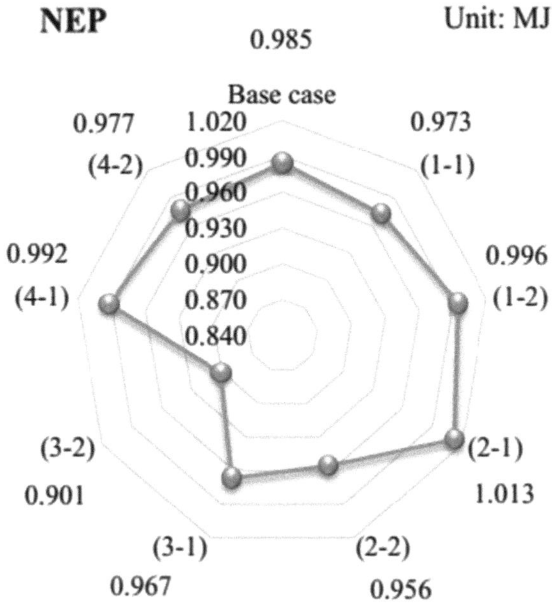

Fig. 5.2 Sensitivity analysis of NEP mitigation

of biogas lamp by 1/6 (a ratio of 7:1:4); (2-2) increasing the powering of biogas lamp by 1/6 (a ratio of 5:5:2); (3-1) decreasing the fueling of a machine by 1/6 (a ratio of 7:4:1); (3-2) increasing the fueling of a machine by 1/6 (a ratio of 5:2:5); (4-1) reducing the cost of biogas equipment by 1/2; (4-2) adding the cost of biogas equipment by 1/2. The results of the sensitivity analysis of energy production are shown in Fig. 5.2. It turns out that the influences of these activities on NEP are disparate in the biogas utilization stage. The substitution of diesel plays an important role by its high energy transforming efficiency. The influence of substituting electricity is also relatively notable, but adjusting the proportion of direct combustion will not significantly change the energy production.

5.3.3 Environmental Benefit Under Different Scenarios

With the joint operation of all five digestate processing pathways as the base case, four scenarios are further developed to determine the optimal energy performance for SIOBS: Scenario 1: substitution of base fertilizer and top dressing; Scenario 2: substitution of base fertilizer and top dressing and forage; Scenario 3: substitution of base fertilizer and top dressing and application of biogas slurry to seed soaking; and Scenario 4: substitution of base fertilizer, top dressing and fertilizer, and application to seed soaking. The energy performance of digestate reuse is indexed by NEP. Regarding a selection in light of energy saving, the solution goes to the base case (i.e., combining all five utilization pathways) due to the significant falloff of NEP in other scenarios compared with the base case (from 29.89 to 109.38%).

5.3.4 Digestate Reuse

The reuse of fermentation digestate is another major application of SIOBS that links different agricultural industries in a more efficient way and replaces the conventional disposal of household wastes (Rehl and Müller 2011). The results show that the substitutions of top dressing and base fertilizers give rise to a significant increase in equivalent energy; i.e., the energy saving due to this process is 9849.299 MJ. The energy consumed for processing and transporting digestate in all the reuse pathways adds up to 13,625.854 MJ. As a result, a negative NEP is discerned in this stage, indicating an energy loss of 226.367 MJ.

5.4 Wastewater Treatment Plants (WWTPs)

5.4.1 Energy Input–Output Analysis

We identified and calculated energy inputs and outputs within the system boundary to address the possibility of net energy production by wastewater treatment combined with biogas–sludge use technologies. The entire life-cycle energy balance was considered, including upstream energy flows (materials extraction, transportation of construction materials, and equipment assembling), on-site energy flows (WWTP operation, maintenance materials, and transportation of sludge), and downstream energy flows (biogas–sludge processing, biogas combustion, sludge incineration, and substitution for industrial fertilizers and pesticides). On this basis, we performed a sensitivity analysis of the energy balance to determine under what circumstances a WWTP can become a net energy producer. The baseline assumption for sludge handling was landfilling, which is the most common sludge handling option in rural areas, although more energy-consuming than the other two options. We carefully examined the impact of variations in gas production efficiency ($\pm$ 10%) on the overall energy balance for the three biogas use alternatives (heat, CHP, and HBU).

5.4.2 System Energy Production

Eight operational scenarios (WS0-WS7), defined in Sect. 4.2.2, were used to determine the optimal performance in terms of environmental impact for different combinations of biogas use and sludge handling.

Figure 5.3 shows the energy consumption/production over all stages of the wastewater treatment system, where different operation scenarios are compared.

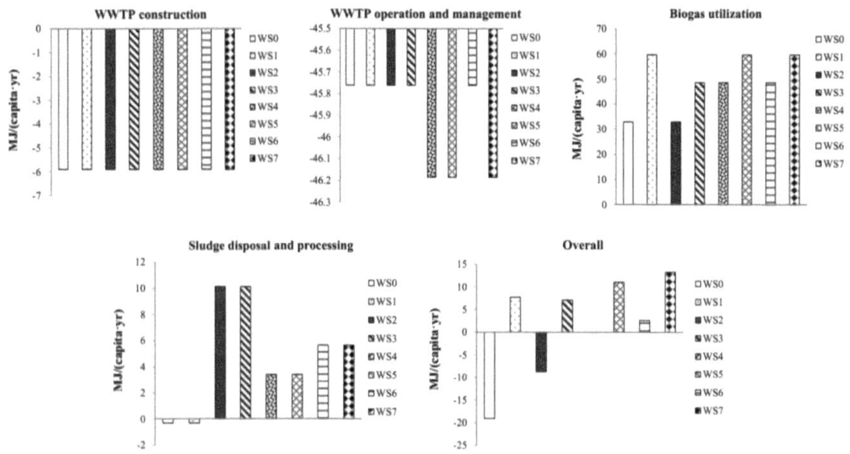

Fig. 5.3 Energy consumption and production during each stage of WWTP under different scenarios

WWTP construction, operation, and maintenance were the energy-consuming processes (a total of 51.66 MJ for the system using landfill or incineration and 52.08 MJ for the system using sludge processing). Daily operation accounted for the greatest consumption (43.49 MJ, > 83%), surpassing all other sources of energy production for the WWTP installation (e.g., 7.25 MJ for construction materials and 0.92 MJ for transportation). Variations among the scenarios resulted in only small changes within these stages, which were primarily due to the different ways of transporting liquid and solid sludge. On the other hand, downstream applications of by-products (biogas and sludge) were major contributors to energy production. Biogas combustion and sludge processing also contributed to energy consumption, either directly or indirectly, including consumption associated with transportation, machinery cost, and fossil fuel combustion. However, the large energy gains from comprehensive biogas use (substitutions for fire coal and thermos electricity) and sludge processing (substitutions for chemical fertilizers and pesticides) resulted in these stages being energy-producing. The biogas use line dominated systemic energy production, where HBU had the highest energy production (59.71 MJ in WS1, WS5, and WS7), followed by CHP (48.65 MJ in WS3, WS4, and WS6) and heat (32.88 MJ in WS0 and WS2). Combined incineration (10.10 MJ in WS2 and WS3) and nutrient processing (3.41 MJ in WS4 and WS5) for handling sludge from anaerobic fermentation were energy-producing (5.64 MJ in WS6 and WS7), while landfilling contributed no system energy savings but incurred costs for vehicle fuels. Overall, in operational scenarios WS1, WS3, WS5, WS6, and WS7, the wastewater treatment system was a net energy producer. In contrast, scenarios WS0 and WS2 still consumed energy, although biogas and sludge were used to some extent. WS4 was almost balanced in energy input–output (−0.02 MJ). Thus, the optimal option for energy performance was WS7 (13.27 MJ), whereas the base scenario (WS0) was the most unsatisfactory choice (−19.08 MJ).

5.4.3 Sensitivity Analysis

As shown in Fig. 5.3, biogas use was the major energy-producing process of the WWTP and the most important factor in achieving net system energy production. How can a wastewater treatment system become a net energy producer? Our results demonstrate that the energy conversion rate in CHP and the biogas yield in HBU were the critical factors in determining system energy performance, while the system energy balance was not sensitive to conversion efficiency when pure heat production was employed (Fig. 5.4). When the heat conversion rate increased by 20% (biogas energy production rising to 39.46 MJ), energy consumption for the wastewater treatment system decreased by 34% compared with the base system, although the system still consumed considerable energy. Provided CHP was in use, the system became a net energy producer when the energy conversion rates were elevated by 10% (58% heat + 32% electricity). When using HBU as the biogas processing pathway, the system had a chance of being a net energy producer even though the biogas production decreased by 10%. Thus, the technologies for external heat and electricity (e.g., heat dissipation and equipment cooling) and biogas production and collection (e.g., temperature and pH control in aerobic tanks) throughout the year were critical in obtaining net energy production.

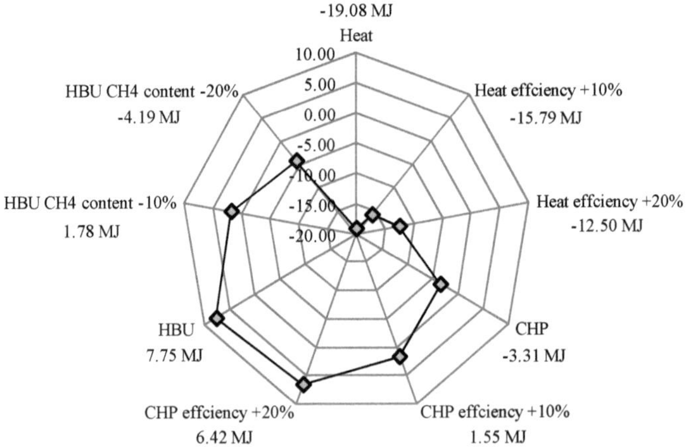

Fig. 5.4 Sensitivity analysis of WWTP energy production. Reprinted from (Chen and Chen 2013b), with permission from Elsevier

References

Chen B, Chen SQ. Life cycle assessment of coupling household biogas production to agricultural industry: a case study of biogas-linked persimmon cultivation and processing system. Energy Policy. 2013a;62(7):707–16.

Chen SQ, Chen B. Net energy production and emissions mitigation of domestic wastewater treatment system: a comparison of different biogas-sludge use alternatives. Biores Technol. 2013b;144(5):296–303.

Chen SQ, Chen B, Song D. Life-cycle energy production and emissions mitigation by comprehensive biogas-digestate utilization. Biores Technol. 2012;114(3):357.

Pöschl M, Ward S, Owende P. Evaluation of energy efficiency of various biogas production and utilization pathways. Appl Energy. 2010;87(11):3305–21.

Rehl T, Müller J. Life cycle assessment of biogas digestate processing technologies. Resour Conserv Recy. 2011;56(1):92–104.

Chapter 6
Economic Assessment of Household Biogas Project

Abstract One of the key concerns of biogas plants is the disposal of comparatively large amounts of digestates in an economically and environmentally sustainable manner. This work analyzes the economic performance (both economic feasibility and economic efficiency) of anaerobic digestion of a given household biogas project (Gongcheng project) based on cost–benefit analysis and DEA method. Firstly, the economic profitability and operation risk were analyzed based on economic indicators. Then, DEA approach was used to measure the relatively economic efficiency of eight household biogas projects in different part of China, based on which the optimization suggestions for economic efficiency can be obtained. Results show that as for the economic feasibility analysis, the net present value (NPV) index as a reflection of the economic feasibility of Gongcheng project is 4754.13 Yuan, showing it has good economic profitability. The payback period (PB) is 2.2 years for the biogas project, which is only 1/5 of the lifetime of the household biogas project (10 years). It can be seen that all the economic investment to support the operation of biogas project could be returned. According to the DEA analysis, the values of technical efficiencies for five biogas projects are smaller than 1 (technical inefficient), indicating that most biogas projects involved are economic inefficient. For further optimization, the input variables ("construction investment" and "maintenance cost") of eight biogas projects should reduce 159.43 (4.55% reduction) and 9.07(5.03% reduction) units (unit: 10^4 yuan), respectively, to increase the economic efficiency. With regard to the output variables, "biogas benefits" variable should increase 33.59 (unit: 10^4 yuan) in total.

Keywords Economic analysis · Cost-–benefit analysis · DEA · Biogas projects

B. Chen et al., *Biogas Systems in China*, DOI 10.1007/978-3-662-55498-2_6

6.1 Economic Feasibility Analysis

6.1.1 System Boundary

The system boundary (see Fig. 6.1) of the biogas project involved contains three phases: construction phase, operation phase, and utilization phase. The first phase covers fundamental preparation activities. Purchased construction materials from outside market are involved for biogas digester construction and installation of gas pipelines and equipment. The fermentation phase is the process of biogas production. Manure and straw are used for fermentation. To keep digester in good use, daily and yearly maintenance activities are also included. Cement and concrete are the materials for repairing and maintenance each year. The last part of utilization shows links between biogas system and other agriculture systems, which is the value added phase of biogas as well as biogas slurry and residue. In this phase, biogas and biogas co-product are in multiuse, and substitute benefits could be gained. Labor and diesel for transportation are also covered in all phases (Chen and Chen 2012; Zhang and Chen 2016a, b).

6.1.2 Economic Costs and Benefits

The substitute economic benefits due to the comprehensive utilization were shown in Fig. 6.2. The total substitute benefits for one household in one year could be

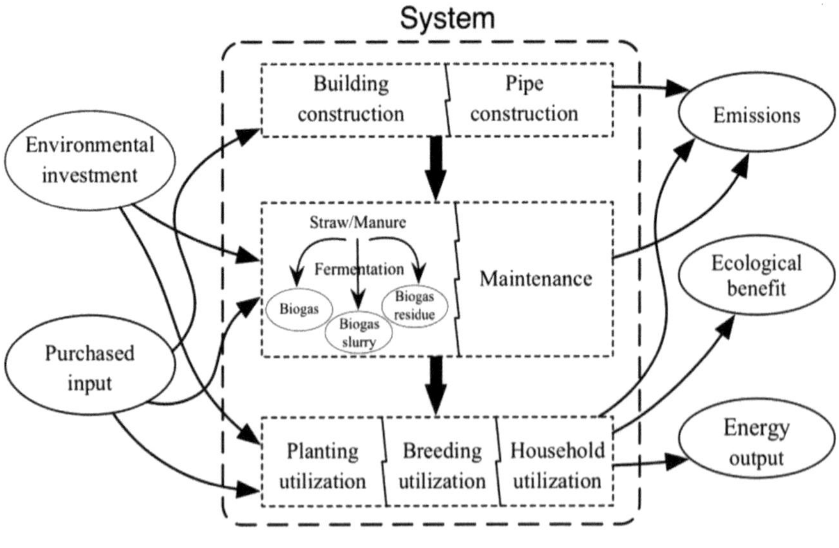

Fig. 6.1 System boundary of biogas project

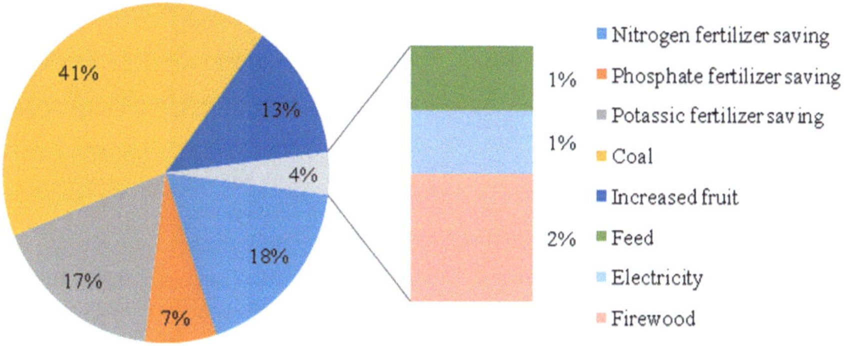

Fig. 6.2 Substitute economic benefit structure of household biogas project

about 1480 Yuan. Figure 6.2 indicates the detailed structure of the economic benefits. Among the economic benefits categories, coal and fertilizer make large proportion, with proportion of 47 and 20%, respectively. The nitrogen fertilizer is the largest contributor in fertilizer alternative benefits, accounting for almost 16%. Increased fruits due to the utilization of biogas slurry and residue make a proportion of 13% in total economic alternative benefits.

The life-cycle stages of biogas project contain construction phase, operation phase (fermentation and maintenance), and utilization phase. Based on the classification of three stages, various of cost flows in every stage are listed and calculated. The economic cost flows in different phases are listed in Table 6.1. The construction cost is equivalent to 1769.43 Yuan, while the rest cost is totally 877 Yuan. It is evident that the construction investments make large proportion in the total inputs. Moreover, labor inputs involved occur in every phase and cover a large percentage during the operation fraction (accounting for 78.6%). Diesel for transporting the construction materials, fermentation materials, and utilization materials makes small proportion as usually household is close to hesperides and the distance between household and paddy field is small.

6.1.3 *Financial Analysis*

Figure 6.3 shows the net present cost and net present benefit in the whole lifespan of biogas project. Based on the two parameters referred above, the indicator net present value (NPV), which is a reflection of the economic feasibility of biogas project, is equivalent to 4754.13 Yuan, showing the biogas project has good economic profitability. The indicator benefit cost ratio (BCR) is equivalent to 1.55. The BCR results of similar studies on biogas projects have a range from 1.08 to 1.39 (Gwavuya et al. 2012; Moreno et al. 2017; Abbas et al. 2017). The efficiency

Table 6.1 Input and output inventory of the household biogas project

Stage	Items	Data	Unit	Unit price	Economic value (Yuan)
Construction stage	Bricks	1320	piece	0.35 Yuan/piece	4.62E+02
	Stone	1	m^3	70 Yuan/m^3	7.00E+01
	Fine sand	1	m^3	50 Yuan/m^3	5.00E+01
	Coarse sand	1.5	m^3	50 Yuan/m^3	7.50E+01
	Cement (425#)	0.95	t	400 Yuan/t	3.80E+02
	Steel	4	m^3	3 Yuan/m^3	1.20E+01
	Plastic mulch	23	m^2	0.74 Yuan/m^2	1.70E+01
	Plastic pipe	2	m	20 Yuan/m	4.00E+01
	Water	950	kg	1.5 Yuan/m^3	1.43E+00
	Transportation	100	km	8.06 Yuan/L	3.62E+02
	Labor	10	person/day	30 Yuan/person	3.00E+02
Operation stage	Straw	94	kg/year	100 Yuan/t	9.40E+00
	Water	5093	kg/year	1.5 Yuan/m^3	7.67E+00
	Excrement	1813	kg/year	/	5.00E+01
	Daily operation labor	0.5	h/day	1 person/time	5.70E+02
	Yearly management	2	time	2 person/time	1.20E+02
	Cement	0.1	t/year	400 Yuan/t	4.00E+01
	Concrete	20	kg/year	390 Yuan/t	7.80E+00
	Transportation	10	km/year	8.06 Yuan/L	3.63E+01
Utilization stage	Transportation	10	km/year	8.06 Yuan/L	3.63E+01
Substitution benefits	Firewood replacement	352	kg	100 Yuan/t	3.52E+01
	Coal	700	kg	1000 Yuan/t	7.00E+02
	Electricity generation	45	kWh	0.45 Yuan/Kwh	2.03E+01
	Increased fruits	150.3	kg	1500 Yuan/t	2.26E+02
	Feed saving	16.4	kg	1080 Yuan/t	1.77E+01
	Nitrogen fertilizer saving	148.02	kg(N)	2.1 Yuan/kg	3.11E+02
	Phosphate fertilizer saving	180	kg(P_2O_5)	0.6 Yuan/kg	1.08E+02
	Potassic fertilizer saving	98.68	kg(K_2O)	3 Yuan/kg	2.96E+02

results show that economic efficiency of the given household biogas project is better than those of most similar biogas projects. Different types of economic investments within the lifespan of biogas project have been effectively payback. The payback period (PB) is 2.2 years for the biogas project, which is only 1/5 of the lifetime of the household biogas project (10 years). It is concluded that all the economic investment to support the operation of biogas project can be returned.

6.2 DEA Analysis

6.2.1 Input and Output Variables

Based on the whole operation process of biogas projects, only two kinds of input variables and three types of output variables are chosen to comprise the data set of DEA method considering the accuracy of the results. In details, the input variables include construction investment, maintenance cost, showing the economic costs status of the biogas project. The output variables contain biogas yield (per day), substitute energy benefit, by-products benefits, revealing the main economic yield of the biogas project. The detailed information about the variables is shown in Table 6.2.

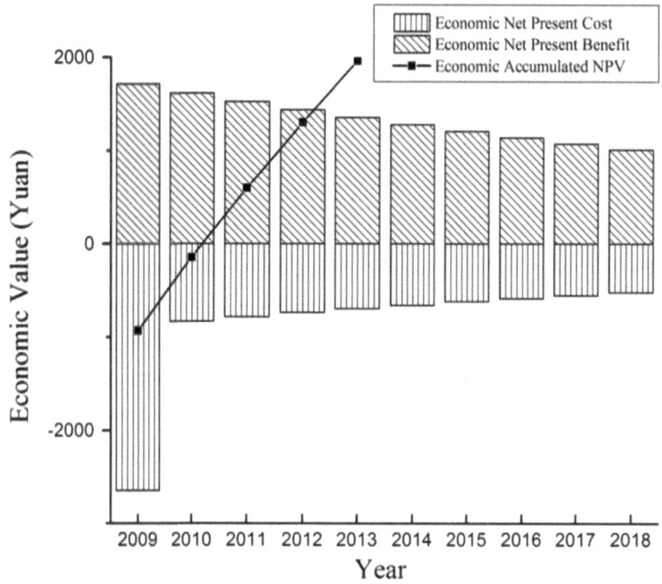

Fig. 6.3 Net present cost and benefit (economic value) of the biogas project from 2009 to 2018

Table 6.2 Selection of input and output variables (one year)

Item	Parameter	Unit	Definition
Input 1	Construction investment (CI)	Yuan (10^4)	Primary cost for the installation
Input 2	Maintenance costs (MC)	Yuan (10^4)/Year	Including the operation and repair costs
Output 1	Biogas yield (Y)	M^3/Year	Biogas generated per year
Output 2	Biogas benefits (BB)	Yuan (10^4)/Year	Substitute benefits for energy
Output 3	By-products benefits (BPB)	Yuan (10^4)/Year	Alternative benefits for fertilizer, etc.

6.2.2 Economic Efficiency Evaluation

Figure 6.4 has shown the results of the DEA analysis of eight similar household biogas projects in different parts of China. According to the DEA model, both technical efficiency, pure technical efficiency, and scale efficiency were calculated. In terms of the pure technical efficiency, five biogas projects have attained the score of 1 (pure technical efficient), and the efficiency of the rest three biogas projects are smaller than 1 (pure technical inefficient). With regard to the scale efficiency, only one biogas project shows positive relationship to its scale and four biogas projects are all negative to their scale. It is notable that almost biogas projects are sensitive to the scale. According to the results of pure technical efficiency and scale efficiency, the values of technical efficiencies for five biogas projects are smaller than 1 (technical inefficient), indicating that most biogas projects involved are economic inefficient.

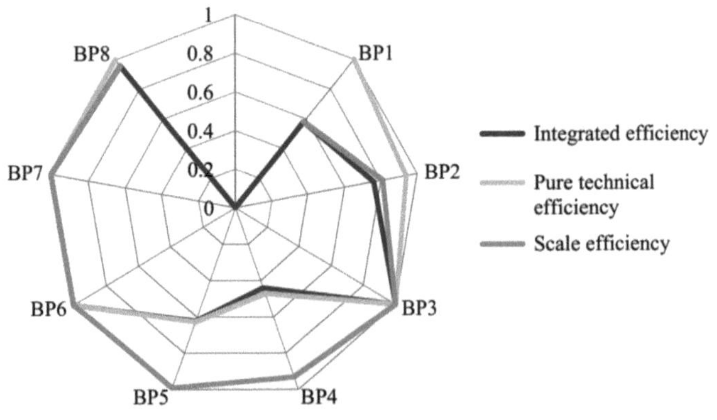

Fig. 6.4 Economic efficiency of biogas projects

6.2.3 Economic Efficiency Optimization

The economic optimization of input and output variables of biogas projects are shown in Table 6.3 according to the DEA results. For the input variables of eight biogas projects, "construction investment" and "maintenance cost" variables will reduce 159.43 (4.55% reduction) and 9.07(5.03% reduction) units (unit: 10^4 yuan), respectively, to increase the economic efficiency. With regard to the output variables, "biogas benefits" variable should increase 33.59 (unit: 10^4 yuan) in total.

As for individual economic efficiency optimization analysis, there are five biogas projects (including Gongcheng biogas project) whose pure technical efficiency values are equivalent to 1, and their specific slack movement as well as radial movement is equivalent to 0, showing that the four biogas projects mentioned above have intelligent economic efficiency from pure technical perspective. With regard to inputs variables, when the value of slacks movement is not equivalent to 0, the investments are redundant and should be mitigated to some extent to increase the integrated economic efficiency. There are three biogas required to mitigate corresponding inputs for the economic efficiency improvement. Taking BP2 as an example, in order to be technical efficient, the amount for construction investment variable should be cut down by 161,600 Yuan, and the maintenance cost variable should be reduced by 14,400 Yuan as well. With regard to outputs variables, when the slacks values are not equivalent to 0, the amount of outputs variables are not redundant and it would be necessary to increase the various yield to make the biogas systems more efficient. In particular, there are three biogas projects need to increase the outputs of different categories of biogas products.

Table 6.3 Optimization of input and output variables of biogas projects

Name	Maintenance cost	Construction investment	Biogas yield	Biogas benefits	By-products benefits
BP1	0	0	0	0	0
BP2	1.44	16.16	1.3	20.49	0
BP3	0	0	0	0	0
BP4	5.26	80.08	0.94	6.54	0
BP5	2.37	63.19	1.42	6.56	0
BP6	0	0	0	0	0
BP7	0	0	0	0	0
BP8	0	0	0	0	0

Note BP1: Liumingying county biogas projects; BP2: Houanding county biogas projects, BP3: Niufang county biogas projects; BP4: Gudian county biogas projects, BP5: Tongying county biogas projects, BP6: Houmiao county biogas projects, BP7: Xilongwan county biogas projects, BP8: Gongcheng biogas project

6.3 Policy Implication

According to the results of DEA, most biogas projects are not economic efficient to achieve the sustainability from the economic perspective. In order to enhance the economic efficiency, suggestions are proposed to improve the integrated economic performance from both input and output perspectives:

(1) Increasing the scale effect. The economic benefits of biogas projects are significant, but the scale effect of biogas project is not clear. Linking biogas projects to the development of local agro-ecosystem could make the largest utilization of biogas project. Developing clean biogas energy, straw, firewood, coal, and other combustion fossil fuel could be replaced, through which greenhouse gas emissions and air pollution could be mitigated, and rural living environment could be enhanced.
(2) Technical parameter upgrading. One of the important reasons for the low efficiency of the current biogas projects is that gas production is not smooth because seasonal volatility affects the total biogas yield. As such, emphases should be put on technologies for the biogas pool construction, fermentation technology, and comprehensive utilization of biogas products (biogas, biogas residue, and biogas slurry). Moreover, technical elements for biogas storage (biogas tank equipment), biogas conversion (biogas power generation and thermal utilization) and other aspects (e.g., biogas battery technology, biogas purification technology, energy crop transgenic technology, biogas power generation technology) should be new breakthrough to overcome technical problems. Also, the establishment of standardized system for biogas industry supporting the upstream and downstream biogas industries should be constructed.
(3) Shifting the development paradigm. With the development of rural economy, commercial energy such as liquefied petroleum gas and natural gas is popularized. Farmers are willing to pay for these convenient energy sources. Due to the fact that price advantage is not obvious of the generated biogas, and maintenance cost is massive, the number of household biogas projects is significantly decreased. Moreover, agriculture is not regarded as the main source of economy revenue for most residents. The young labors generally go out for work, and the rest residents are mainly the elderly, women, and children, who have small demand for energy. Under this situation, it would be better to develop centralized biogas supply project based on local actual situation, while investment in household biogas projects should be reduced and transfer to the large- and medium-sized biogas projects. In addition, it should be systematically made to develop different biogas models to consider both quality and quantity of biogas projects.
(4) Fermentation materials upgrading. Focus on the development and promotion of energy crop served as fermentation materials of biogas project. As the largest agricultural country in the world, there are abundant straw resources in China. However, less than one-tenth of the total available straw is utilized, leading to

large amount of biomass resource waste. Compared with the manure-based biogas projects and industrial waste-based biogas projects, the biogas yield could be enhanced for straw-based biogas projects. In addition, energy crop (corn, beets, perennial) could be cultivated to replace petrochemical raw materials to improve gas production.

References

Abbas T, Ali G, Adil SA, Bashir MK, Kamran MA. Economic analysis of biogas adoption technology by rural farmers: the case of Faisalabad district in Pakistan. Renew Energy. 2017;107:431–9.

Chen SQ, Chen B. Sustainability and future alternatives of biogas-linked agrosystem (BLAS) in China: an emergy synthesis. Renew Sustain Energy Rev. 2012;16(6):3948–59.

Gwavuya SG, Abele S, Barfuss I, Zeller M, Müller J. Household energy economics in rural Ethiopia: a cost-benefit analysis of biogas energy. Renew Energy. 2012;48(6):202–9.

Moreno L, González A, Cuadros-Salcedo F, Cuadros-Blázquez F. Feasibility of a novel use for agroindustrial biogas. J Clean Prod. 2017;144:48–56.

Zhang BY, Chen B. Dynamic hybrid life cycle assessment of CO_2 emissions of a typical biogas project. Energy Procedia. 2016a;104:396–401.

Zhang BY, Chen B. Sustainability accounting of a household biogas project based on emergy. Appl Energy. 2016b;194:819–31.

Chapter 7
Emergy Analysis of Biogas-Linked Agricultural System

Abstract This chapter provides an overview of the economic and environmental performance of biogas-linked agricultural system (BLAS) in China based on emergy. A set of emergy indices are incorporated to describe the energy and materials transformation within the system, and an emergy-based CO_2 emission indicator ($EmCO_2$) is proposed to achieve low-carbon optimization of the whole system. Emergy synthesis and emergetic ternary diagram are then utilized to evaluate the BLAS and its subsystems, with scenario analysis performed to identify a more sustainable development pathway for the BLAS. Finally, a framework is developed to track dynamical behaviors of the whole system (Level I), transforming process (Level II), and resource component (Level III) simultaneously, and two new indicators, emergy contribution rate (ECR) and emergy supply efficiency (ESE) are proposed to address the contribution and efficiency of resource components within each process. The results showed that BLAS made a favorable contribution to carbon mitigation and was more environment-friendly than the traditional agricultural systems. Scenario analysis demonstrated that continual biogas construction and effective technological revolution were preferable routes to further improve the whole system's performance. It can be concluded that breeding and biogas subsystems were economic input-dependent. Electricity and diesels were the most efficient components in supplying all the processes in BLAS. The relatively high transformities and the constant descent of sustainability within all processes are the key problem that hinders the promotion of BLAS.

Keywords Biogas-linked agricultural system · Emergy analysis · CO_2 emission · Ternary diagram · Sustainability · Efficiency

7.1 Emergy Analysis

Although there have been some researches on the sustainability evaluation of biogas projects, analyses on the overall performance and inner interactions of the biogas-linked agricultural system (BLAS) are still few. Moreover, since biogas

B. Chen et al., *Biogas Systems in China*, DOI 10.1007/978-3-662-55498-2_7

project as a kind of renewable energy engineering has been widely used in rural areas, its carbon mitigation performance cannot be ignored. A carbon budget of the concerned system is necessary to be highlighted associated with the conventional emergy analysis. In this chapter, the structure, function, efficiency, and sustainability of a BLAS are analyzed through emergy analysis from a systematic perspective. The carbon mitigation performance of the biogas production is also evaluated via an emergy-based CO_2 emission indicator.

7.1.1 Methodology

(1) **Emergy diagram**

As the activities in biosphere are usually considered to be driven by solar energy, different forms of energy are assumed to be derived or transformed from solar energy directly or indirectly. Thus, emergy as abbreviation of embodied solar energy, now, is accepted as a common approach to study all kinds of system containing different kinds of energy or material (Campbell 1998; Jiang et al. 2007; Odum 1996; Yang and Chen 2014). Through emergy analysis, each form of energy carrier in the concerned system is translated into its embodied solar energy equivalent, or solar emergy, by its solar transformities as a conversion factor. Solar transformity (solar energy joules per unit energy joule in unit of sej/J) is used to calculate the quality of energy or resource and its position in the universal energy transformation hierarchy. The larger the transformity is, the more solar energy required for the production and maintenance of the resource, product, or service, and the higher their position in the energy hierarchy of the universe (Odum 1988, 1996; Chen and Chen 2014).

The emergy diagram of a typical BLAS in Gongcheng County of China is depicted in Fig. 7.1. The emergy inputs can be divided into two categories: (1) the emergy obtained freely from the environment (including renewable resources from the free environment such as sunlight, wind, kinetic energy, earth cycle, and nonrenewable resources of soil loss) and (2) the auxiliary artificial emergy: nonrenewable purchased emergy (chemical fertilizer, fuel, pesticide, electricity, and building materials) and renewable organic emergy (labor, seeds, organic fertilizer, feeds). The emergy outputs of this coproduction system comprise agricultural products (rice, bean, peanut, corn, fruit, vegetables), breeding outputs (pigs, fishes), and biogas and its by-products (biogas slurry and residue).

(2) **Emergy indices**

In addition to the conventional emergy indices (Odum 1996; Brown and Ulgiati 1997), such as emergy yield ratio (EYR), environmental loading ratio (ELR), emergy investment ratio (EIR), and emergy sustainable index (ESI), some indicators that reflect the performance and inner interactions of the system are also

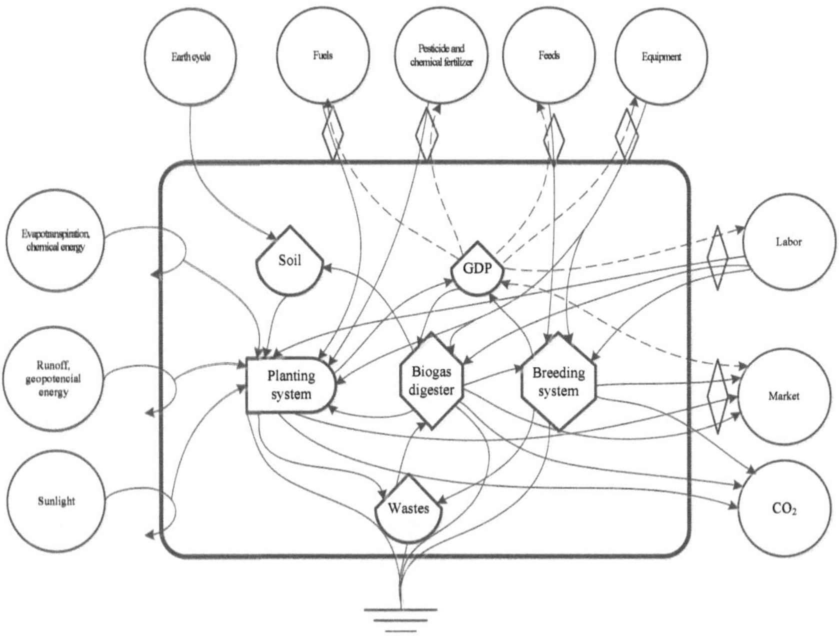

Fig. 7.1 Emergy diagram of the BLAS. Reprinted from Yang and Chen (2014), with permission from Elsevier

developed and demonstrated in Eqs. (7.1)–(7.6). The system-level diversity ratio (SDR), derived from the modified Shannon information formula, provides a quantitative assessment of the systematic diversity, which represents the ratio of actual diversity to the maximum potential diversity of a specific system (Ulgiati et al. 2011). Thus, if SD is closer to SD_{max} (i.e., SD/SD_{max} is closer to 1), it implies a higher system resilience.

$$SD = -\sum\left[\left(\frac{Y_i}{Y}\right)\right]\ln\left(\frac{Y_i}{Y}\right) \tag{7.1}$$

$$SDR = SD/SD_{max} \tag{7.2}$$

where Y_i = emergy output of the ith flow = (amount of ith flow; J or g) × (emergy intensity of the ith flow; sej/J or sej/g). Y is the total yield of the system including all kinds of flows in the system, SD is the system diversity, and SD_{max} is the maximum potential diversity when the total emergy is evenly assigned to each output.

As an illustration of the complexity of emergy yield, system dominance degree index (SDI) is also employed. A larger SDI indicates that the emergy is distributed unevenly, and the role of subsystem plays in the whole system.

$$\mathrm{SDI} = \sum \left(\frac{Y_i}{Y}\right)^2 \tag{7.3}$$

Besides, the indicators of self-support ratio (SSR) and self-support orientation (SSO) are employed to monitor the internal operation mechanism (Zhang et al. 2007a, b). In the BLAS, biogas slurry and residue can be used as the replacements of traditional chemical fertilizer for planting industry and the feed for breeding subsystem. Specifically, the planting and breeding subsystems provide raw materials for anaerobic digestion of biogas production. However, the system analysis of the whole BLAS can hardly explicitly explore the internal interactions of subsystems. The indicators of SSR and SSO, which can reflect the internal interactions between subsystems and trace the inter-subsystem material flows, are given as below:

$$\mathrm{SSR} = \frac{I_{b-i} + I_{i-b}}{\mathrm{RP_{biogas}} + \mathrm{RP}_i} \tag{7.4}$$

where SSR is the self-support ratio, i indicates the specific subsystem, $\mathrm{RP_{biogas}}$, and RP_i denote the total purchased renewable emergy of biogas subsystem and other subsystems, respectively. I_{b-i} is the flow from biogas subsystem to other subsystems, while I_{i-b} is the opposite.

$$\mathrm{SSO} = \frac{I_{b-i} - I_{i-b}}{I_{b-i} + I_{i-b}} \tag{7.5}$$

where SSO is the self-support orientation index specifying the self-support orientation of the systematic internal interactions. If the index value is −1, crop production (or breeding) would have no support from biogas system, while if it is 1, biogas digestion would have no support from other subsystems. If it is 0, the mutual support emergy flux will be equal to each other.

Also, as biogas is a competitive resource compared with other energy sources in terms of resources saving and carbon reduction, the carbon reduction function of the biogas system as an important contributor to the sustainable development should be emphasized. Therefore, $E_m\mathrm{CO_2}$, as an emergy-based CO_2 emission indicator, is put forward to present a new goal function for possible low-carbon system optimization. As shown in Eq. (7.6), $E_m\mathrm{CO_2}$ represents the ratio of CO_2 emission during the entire construction and operation stage of biogas project. The lower the $E_m\mathrm{CO_2}$, the better the low-carbon optimization degree based on sustainability framework.

$$E_m\mathrm{CO_2} = C/Y/\mathrm{ESI} \tag{7.6}$$

where C is the total CO_2 emission in the biogas production system, Y is the yield of biogas with the unit of sej, and ESI is the sustainability of the biogas system that is determined by EYR and ELR. Obviously, the larger the CO_2 emission per emergy output, the higher the E_mCO_2. On the contrary, the lower the system sustainability, the higher the E_mCO_2.

The indicators are summarized and listed in Table 7.1.

Table 7.1 Emergy-based indices and ratios for the BLAS

Items	Expressions	References	Implications
Total emergy inputs (T)	RR + RP + NR + NP	Odum (1996)	Total emergy flux in a system
The proportion of renewable emergy inputs (R)	$(\mathrm{RR}+\mathrm{RP})/T$	Odum (1996)	The dependence of the system on renewable emergy
The proportion of purchased emergy inputs (P)	$(\mathrm{RP}+\mathrm{NP})/T$	Odum (1996)	The dependence of the system on purchased emergy
Emergy investment ratio (EIR)	(NP + RP)/(RR + NR)	Brown and Ulgiati (1997)	The ratio of emergy from outside the system to the indigenous emergy inputs
Emergy yield ratio (EYR)	$Y/(\mathrm{NP}+\mathrm{RP})$	Brown and Ulgiati (1997)	Economic efficiency of BLAS
Environmental loading ratio (ELR)	(NR + NP)/(RR + RP)	Brown and Ulgiati (1997)	Environmental loading exerted by the BLAS
Emergy sustainable index (ESI)	EYR/ELR	Brown and Ulgiati (1997)	Sustainability of the BLAS
System-level diversity index (SDR)	$-\sum\left[\left(\frac{Y_i}{Y}\right)\right]\ln\left(\frac{Y_i}{Y}\right)$	Brown et al. (2006)	The diversity of the BLAS
System dominance degree index (SDI)	$\sum\left(\frac{Y_i}{Y}\right)^2$	Lan et al. (2002)	The complexity of the BLAS
Self-support ratio (SSR)	$\frac{I_{b-i}+I_{i-b}}{\mathrm{RP}_{\mathrm{biogas}}+\mathrm{RP}_i}$	Zhang et al. (2007a, b)	The inner flows between subsystems
Self-support orientation (SSO)	$\frac{I_{b-i}-I_{i-b}}{I_{b-i}+I_{i-b}}$	Zhang et al. (2007a, b)	The direction of inner material flow
E_mCO_2	$E_mCO_2 = C/Y/\mathrm{ESI}$	This paper	The ratio of real CO_2 released to the emergy-based sustainability indicator per sej biogas yield

7.1.2 Results and Discussion

i. **Evaluation of the BLAS**

In Fig. 7.2, the total emergy input of BLAS is 1.80E+21 sej. The proportions of renewable emergy from the environment, nonrenewable emergy from the environment, nonrenewable purchased emergy, and renewable purchased emergy to the total emergy input are 16.8, 1.1, 71.4, and 10.7%, respectively. The nonrenewable purchased emergy inputs come from electricity, chemical fertilizer, and equipment used for planting and breeding. Fertilizer input plays a fundamental role in this system. In the future evolution of agricultural system in Gongcheng, how to reduce the utility of chemical fertilizer and rationally make use of biogas slurry and residue as substitution will be the key to realizing sustainable development. The nonrenewable emergy from the environment constitutes 1.1% of the total emergy, indicating that soil is of little significance to the yield of BLAS.

The emergy indices associated with sustainability are presented in Table 7.2.

Comparisons of BLAS with the other three agricultural systems are depicted in Fig. 7.3. EIR of BLAS (4.58) is higher than the average value of Chinese agriculture, indicating that it depends on purchased emergy inputs. However, this value is much less than that of Italian agricultural system in 1989 (8.52), reflecting that although the BLAS's modernization level is on a relatively high compared with that of Chinese average level, the BLAS of Gongcheng still needs further development.

The EYR depicts the potential contribution of the agricultural system to the total economy. In an economic view, the output of a system should not be larger than the investment. The higher the EYR, the higher the system yields per unit of input emergy, and the greater the contribution of the purchased (renewable and nonrenewable) emergy to the whole system (Bastianoni and Marchettini 2000). As shown in Fig. 7.3a, the EYR of the BLAS in the year of 2009 was only 0.63, much lower than 2.08 of the whole China, 1.12 of Italy, and 0.72 of the "Four-in-One" peach

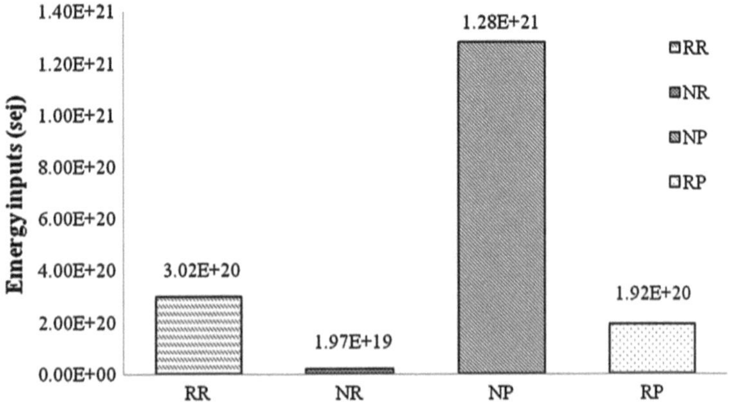

Fig. 7.2 Emergy inputs of the BLAS

Table 7.2 Selected results of emergy analysis of the BLAS

Items	Expressions	Values
EIR	(NP + RP)/(RR + NR)	4.58
EYR	$Y/(\mathrm{NP}+\mathrm{RP})$	0.63
ELR	$(NR+\mathrm{NP})/(\mathrm{RR}+\mathrm{RP})$	2.64
ESI	EYR/ELR	0.24
SDR	$-\sum\left[\left(\frac{Y_i}{Y}\right)\right]\ln\left(\frac{Y_i}{Y}\right)$	0.35
SDI	$\sum\left(\frac{Y_i}{Y}\right)^2$	0.56

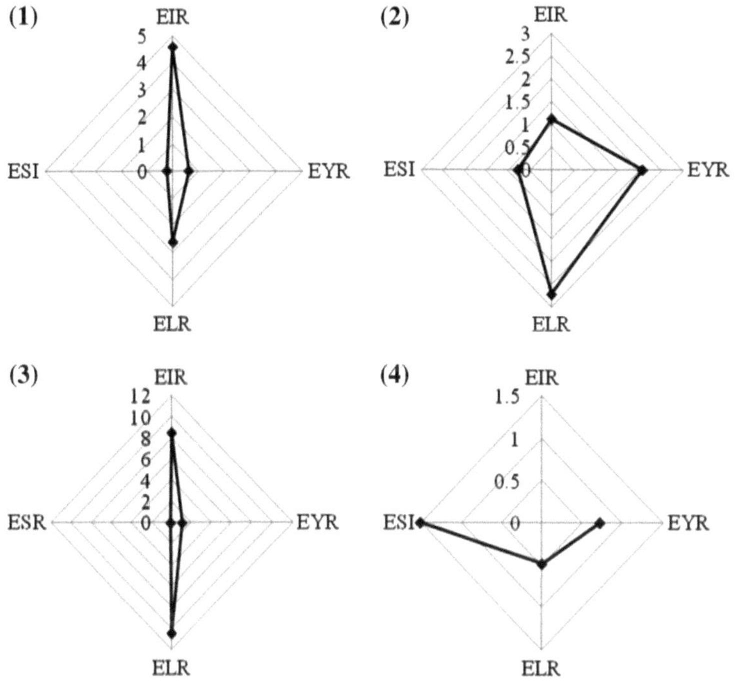

Fig. 7.3 Comparisons of emergy indices of different agricultural systems (*1* BLAS in this paper, *2* Chinese agricultural system in 2000, *3* Italian agricultural system in 2000, *4* a "Four-in-One" peach production system in Beijing)

production system. The lowest EYR of this agricultural system should be attributed to the backward economic and mechanized development level in minority inhabited areas, manifested as follows: (1) The singular planting and breeding model still predominates the whole agriculture system, which makes most farmers rely on the traditional crops and livestock (rice, corn, pig); (2) since there are few trades among different communities, the community is still at self-sufficient stage; (3) almost no added values are produced by the agricultural production, exchanges with the outside being focusing on the primary products.

The ELR is related to the fraction of renewable resources, which can be regarded as a measure of ecosystem stress due to production (Ulgiati and Brown 1998). The ELR of the BLAS is smaller (2.63) compared with that of the whole China and Italy. Due to inner recycling of the material and energy in the whole system (chemical fertilizer is substituted by organic fertilizer of biogas slurry and residue), it avoids purchasing them from outside. Thus, environmental pressure is alleviated. However, the ELR of the BLAS is higher than that of the typical "Four-in-One" peach production system, indicating the inner recycling has room to be further accelerated and promoted.

ESI relates to both EYR and ELR. The value of ESI of Gongcheng is larger than that of Italy, but smaller than that of the average China and the "Four-in-One" peach production system. Although the environmental stress on the agricultural system of Gongcheng is relatively slight, EYR is weak for the whole assessment; i.e., the economic contribution needs further promotion for local economy. In summary, the improvement of the production efficiency as well as optimization of the material recycling within system is the key to the development of BLAS.

SDR represents the diversity of one system. A higher SDR (closer to 1) means more resilience when fluctuations occur. The SDR of the BLAS is only 0.35, indicating that the BLAS's production structure is too simple, and diversification of agricultural system should be fostered in the future.

SDI depicts the balance degree of subproduction systems of an agricultural system. Without a high proportion of the breeding and the biogas yields, the SDI of the BLAS is only 0.56.

ii. **Emergy analysis of subsystems**

(1) **Planting subsystem**

As listed in Table 7.3, in the planting subsystem, the proportions of RR, NR, NP, and RP are 17.3, 1.12, 70.1, and 11.5%, respectively. As for RR, the rainfall accounts the most. Soil losses' portion is negligible, which indicates that the climate and soil quality are in a good status in this area. It is obvious that NP contributes the most, especially the usage of chemical fertilizer constituting 77% of the total NP. However, organic fertilizer derived from the biogas residue and slurry still plays an important role in the substitution of traditional chemical fertilizer, which accounts for 34.3% of the total RR from economy. In all kinds of yields in the planting subsystem, fruit production occupies the largest proportion and takes the dominant position.

The value of EYR (0.47) is much lower than the national average, indicating the low efficiency of usage of local resources in planting subsystem. The EIR is 4.44 for crop production, indicating that a large amount of purchased emergy is put into the system, and the value of ELR is 2.47, less than 2.72 of the national average. However, ESI is also much lower than the national average due to the low EYR.

Table 7.3 Emergy analysis of planting subsystem

Items	Data	Unit	Solar transformity	Unit	References	Solar emergy (sej)
Renewable emergy from the environment (RR)						
Sunlight	2.71E+18	J	1.00	sej/J	Odum (1996)	2.71E+18
Wind, kinetic energy	2.51E+07	J	2.45E+03	sej/J	Odum (2000)	6.15E+10
Evapotranspiration, chemical energy	4.57E+15	J	3.02E+04	sej/J	Odum (1996)	1.38E+20
Runoff, geopotential energy	6.8E+15	J	1.76 E+04	sej/J	Odum (2000)	1.20E+20
Earth cycle	6.43E+14	J	5.80 E+04	sej/J	Odum (2000)	3.73E+19
Subtotal						2.98E+20
Nonrenewable emergy from the environment (NR)						
Net loss of topsoil	2.749E+11	J	2.08 E+05	sej/J	Odum (1996)	1.94E+19
Nonrenewable purchased emergy (NP)						
Nitrogenous fertilizer	2.56E+10	g	2.41E+10	sej/g	Brandt-Williams (2002)	6.16E+20
Phosphate fertilizer	1.15E+10	g	2.20E+10	sej/g	Brandt-Williams (2002)	2.53E+20
Potash fertilizer	9.03E+09	g	1.74E+09	sej/g	Brandt-Williams (2002)	1.57E+19
Compound fertilizer	1.80E+10	g	2.80E+09	sej/g	Lan et al. (2002)	5.03E+19
Pesticide	7.35E+08	g	1.48E+10	sej/g	Brandt-Williams (2002)	1.09E+19
Diesel	4.25E+14	J	1.11 E+05	sej/J	Odum (1996)	4.72E+19
Agricultural equipment depreciation	4.15E+09	g	6.70E+09	sej/g	Brown and Arding (1991)	2.78E+19
Agricultural film	5.88E+10	g	3.20E+09	sej/g	Brown and Bardi (2001)	1.88E+20
Agricultural electricity	4.07E+11	J	2.69 E+05	sej/J	Odum (1996)	1.09E+17
Subtotal						1.21E+21
Renewable purchased emergy (RP)						
Sapling	4.81 E+06	J	4.94E+12	sej/$	Wei et al. (2009)	2.38E+19
Organic fertilizer	2.51E+15	J	2.70 E+04	sej/J	Wei et al. (2009)	6.78E+19

(continued)

Table 7.3 (continued)

Items	Data	Unit	Solar transformity	Unit	References	Solar emergy (sej)
Farming labor	2.59E +14	J	3.80 E+05	sej/J	Lan et al. (2002)	9.84E+19
Seeds	1.01E +14	g	7.86 E+04	sej/J	Cohen et al. (2006)	7.97E+18
Subtotal						1.98E+20
Total input (T)					1.72E+21	
Yield (Y)						
Rice	6.94E +14	J	8.70 E+04	sej/J	Ulgiati et al. (1993)	6.04E+19
Beans	5.20E +13	J	1.29 E+05	sej/J	Ulgiati et al. (1993)	6.69E+18
Corn	2.24E +14	J	1.01 E+05	sej/J	Ulgiati et al. (1993)	2.27E+19
Peanuts	1.93E +12	J	1.30 E+05	sej/J	Ulgiati et al. (1993)	2.50E+17
Vegetables	1.38E +14	J	2.70 E+04	sej/J	Lan et al. (2002)	3.73E+18
Fruits	1.51E +15	J	3.81 E+05	sej/J	Ulgiati et al. (1993)	5.76E+20
Total output					6.70E+20	

(2) **Breeding subsystem**

According to Table 7.4, the proportions of RR, NR, NP, and RP are 1.57, 0.17, 23.6, and 74.6%, respectively. RP constitutes the largest proportion. In this subsystem, as feed for pig breeding and aquaculture, biogas slurry is largely used, accounting for 36.2% of the total inputs.

Compared with the planting subsystem, the EYR (0.78) in the breeding subsystem is higher, which indicates that the breeding subsystem depends less on the purchased emergy from outside. Owning to a large amount of renewable purchased emergy input, the EIR for livestock production (0.31) is much lower than the planting subsystem and the national average level (2.72), indicating that the breeding subsystem in BLAS is under slight environmental pressure. Correspondingly, ESI of the breeding subsystem is also higher than the planting subsystem due to higher EYR and lower ELR.

(3) **Biogas subsystem**

As shown in Table 7.5, the total emergy input to the biogas system amounts to 3.28E+20 sej/yr, of which the RN from the environment constitutes only a small fraction. RP inputs and NP inputs contribute 88.8 and 10.8%, respectively. The local renewable emergy inputs make up the highest proportion in this system.

Table 7.4 Emergy analysis of breeding subsystem

Items	Data	Unit	Solar transformity	Unit	References	Solar emergy (sej)
Renewable emergy from the environment (RR)						
Sunlight	3.84E+16	J	1.00E+00	sej/J	Odum (1996)	3.84E+16
Wind, kinetic energy	3.55E+05	J	2.45E+03	sej/J	Odum (2000)	8.71E+08
Evapotranspiration, chemical energy	6.47E+13	J	3.02E+04	sej/J	Odum (1996)	1.95E+18
Runoff, geopotential energy	9.63E+07	J	1.76E+04	sej/J	Odum (2000)	1.69E+12
Earth cycle	9.10E+12	J	5.80E+04	sej/J	Odum (2000)	5.28E+17
Subtotal						2.52E+18
Nonrenewable emergy from the environment (NR)						
Net loss of topsoil	1.32E+12	J	2.08E+05	sej/J	Odum (1996)	2.74E+17
Nonrenewable purchased emergy (NP)						
Breeding electricity	4.56E+10	J	2.69E+05	sej/J	Odum (1996)	1.22E+16
Building materials of breeding	1.92E+10	g	1.97E+09	sej/g	Brown and Bardi (2001)	3.78E+19
Subtotal						3.78E+19
Renewable purchased emergy (RP)						
Breeding labor	2.78E+12		3.80E+05	sej/J	Lan et al. (2002)	1.06E+18
Fingerling	4.65E+05	$	3.70E+12	sej/$	Coelho et al. (2003)	1.72E+18
Feed	2.10E+10	g	2.80E+09	sej/g	Lan et al. (2002)	5.88E+19
Biogas slurry	2.15E+15	J	27,000	sej/J	Wei et al. (2009)	5.80E+19
Subtotal						1.20E+20
Total inputs (U)						1.60E+20
Yield (Y)						
Pork	3.81E+14	J	3.36E+06	sej/J	Odum and Odum (1983)	1.22E+20
Fishery products	2.56E+11	J	3.36E+06	sej/J	Odum (1996)	8.59E+17
Total yield						1.23E+20

Specifically, human labor contributes a great proportion, indicating that the biogas production has great potential to create rural employment opportunities for surplus labors.

The emergy yield of the biogas project includes biogas, biogas slurry, and biogas residue, which amounts to 2.10E+20 sej/yr. Biogas as the major product makes up 62.86% of the total emergy yield (1.32E+20 sej/yr).

Table 7.5 Emergy inputs and outputs of biogas subsystem

Items	Data	Unit	Solar transformity	Unit	References	Solar emergy (sej)
Renewable emergy from the environment (RR)						
Sunlight	1.65E+15	J	1.00E+00	sej/J	Odum (1996)	1.65E+15
Wind, kinetic energy	1.52E+04	J	2.45E+03	sej/J	Odum (2000)	3.73E+07
Evapotranspiration, chemical energy	2.77E+12	J	3.02E+04	sej/J	Odum (1996)	8.38E+16
Runoff, geopotential energy	4.13E+12	J	1.76E+04	sej/J	Odum (2000)	7.26E+16
Earth cycle	3.90E+11	J	5.80E+04	sej/J	Odum (2000)	2.26E+16
Subtotal						1.81E+17
Nonrenewable emergy from the environment (NR)						
Net loss of topsoil	6.78E+12	J	2.08E+05	sej/J	Odum (1996)	1.41E+18
Nonrenewable purchased emergy (NR)						
Sands	1.46E+10	g	1.97E+09	sej/g	Brown and Bardi (2001)	2.88E+19
Steel	3.18E+07	g	4.13E+09	sej/g	Wei et al. (2009)	1.31E+17
Plastic film	9.16E+04	g	3.20E+09	sej/g	Brown and Bardi (2001)	2.93E+14
Cement input	3.18E+09	g	1.97E+09	sej/g	Wei et al. (2009)	6.26E+18
Diesel	1.96E+12	J	1.06E+05	sej/J	Odum (2000)	2.08E+17
Subtotal						3.54E+19
Renewable purchased emergy (RP)						
Dropped fruits and straws	1.45E+15	J	1.49E+05	sej/J	Brown and Ulgiati (2001)	2.17E+20
Animal manure	6.97E+14	J	2.70E+04	sej/J	Wei et al. (2009)	1.88E+19
Biogas maintenance labor	1.47E+14		3.80E+05	sej/J	Lan et al. (2002)	5.58E+19
Subtotal						2.91E+20
Total inputs (U)						3.28E+20
Yield (Y)						
Biogas	5.32E+14	J	2.48E+05	sej/J	Zhou et al. (2010)	1.32E+20
Biogas slurry and residue	2.90E+15	J	2.70E+04	sej/J	Wei et al. (2009)	7.84E+19
Total yield						2.10E+20

EYR is 0.64, higher than that of the planting subsystem but lower than that of the breeding subsystem. ELR below 2 indicates relatively low environmental impacts (Cavalett et al. 2006). The ELR is only 0.13, much lower than those of the other systems, implying that biogas project puts little pressure on the environment owning to a large proportion of RR and a small percentage of PN input. The ESI is 5.13, showing that the biogas project has higher utilization efficiency and lower environmental loads, implying an energy-saving and environment-friendly ecological project.

As a substitution of traditional fossil fuels, biogas can avoid CO_2 emissions from coal and direct biomass combustion. The amount of emission reduction by biogas is calculated and demonstrated in Table 7.6. In South China, the household biogas yield of 450 m^3 can replace 700 kg coal combustion and reduce 1500 kg straws for direct combustion. In Table 7.6, a negative value means it is carbon-reducing, while a positive value means it is carbon-releasing. Assuming that the density of wood is 1.54 kg/m^3 and the average forestry storage is 90 m^3/ha, the result shows that the utility of biogas would avoid 2.82t CO_2 emissions per household and increase 2700 m^3 annual forestry storage in Gongcheng. Thus, it plays an indispensable role in the mitigation of CO_2 concentration.

To figure out whether biogas is a carbon-reducing renewable energy source, the total CO_2 emissions including the direct and the indirect ones that are embodied in the input materials should be taken into account. Considering CO_2 released in the transportation, construction, and operation stages of the biogas engineering, the CO_2 emission intensity as well as E_mCO_2 is depicted in Table 7.7. A lower CO_2 emission intensity indicates that less CO_2 is emitted when 1 J of biogas is produced. A lower E_mCO_2 of biogas production is observed due to lower CO_2 emission intensity and higher sustainability. The CO_2 emission intensity of the biogas system is

Table 7.6 Carbon emission reduction potential of each biogas digester per year in Gongcheng

Items	Values	CO_2 emission change/ton
Substitution of coal	700 kg	−1.04
Substitution of firewood	352 kg	−0.47
Substitution of straw	1500 kg	−1.87
Biogas yield	450 m^3	+0.53
Total	–	−2.82

Table 7.7 Comparison of CO_2 emission indicators for different biomass utility alternatives

Indicators	Unit	Biogas system	JCL oil-based biodiesel	Wheat-based bioethanol	Corn-based bioethanol
CO_2 emission intensity	g/J	2.71E-06	4.26E-05	6.99E-05	6.38E-05
E_mCO_2	g/sej	5.71E-11	2.97E-10	1.01E-09	2.44E-09

Table 7.8 Comparison of emergy-based indices of three subsystems

Items	Expressions	Cropping subsystem	Breeding subsystem	Biogas subsystem
EYR	$Y/(\mathrm{NP}+\mathrm{RP})$	0.48	0.78	0.64
ELR	$(\mathrm{NR}+\mathrm{NP})/(\mathrm{RR}+\mathrm{RP})$	2.48	0.31	0.13
ESI	EYR/ELR	0.19	2.51	5.10
SSR	$\frac{I_{b-i}+I_{i-b}}{\mathrm{RP}_{\mathrm{biogas}}+\mathrm{RP}_i}$	0.14	0.19	
SSO	$\frac{I_{b-i}-I_{i-b}}{I_{b-i}+I_{i-b}}$	1	0.51	

2.71×10^{-6} (g CO_2/J), much lower than 6.99×10^{-5} (g CO_2/J) of wheat (Dong et al. 2008), 6.38×10^{-5} (g CO_2/J) of corn (Giampietro and Ulgiati 2005), and 4.26×10^{-5} (g CO_2/J) of Jatropha curcas L. production system (Ju and Chen 2011), indicating that compared with other biomass utility alternatives, biogas is more sustainable under the same CO_2 emission and has a great carbon reduction potential.

(4) **Comparisons of emergy indicators of three subsystems**

The comparison of emergy-based indices calculated for subsystems of BLAS is presented in Table 7.8. The breeding subsystem has the largest EYR, indicating a relatively higher contribution of the local (RR and NR) sources of emergy to the system; i.e., the breeding subsystem makes the largest contribution to the economic development of Gongcheng. The ELR indicates that the high environmental loading of the cropping subsystem is exerted by the extensive consumption of chemical fertilizers. Biogas subsystem has the lowest environmental pressure owning to a large proportion of renewable purchased emergy input, such as human labor. As a result, biogas subsystem has the highest sustainability compared with another two subsystems, indicating that biogas system can be regarded as an environmentally friendly and sustainable energy project.

SSR is used to quantify the flows among different subsystems. In Table 7.8, SSR indicates that the exchange fluxes between biogas system and the planting system (breeding system) account for 14% (19%) of the total renewable purchased resources. Between biogas and planting subsystems, SSO equal to 1 shows that direction of material flow is from the biogas to the planting system. In addition, the value of 0.5 between biogas and breeding subsystems further supports the conclusion that the direction of emergy fluxes is from biogas subsystems to the other ones. Thus, biogas digester can produce by-products such as biogas slurry and residue that are useful for breeding and planting as manure and straws.

7.1.3 Concluding Remarks

Emergy analysis as a useful tool can measure the sustainability level of compound agricultural system. Based on the analysis, the conclusions are offered as below:

The biogas production process is found to be carbon-reducing. Further promotion of BLAS mode in rural areas is consistent with the national objectives of "sustainable rural development" and "low-carbon society."

Due to the stronger recycling of biogas slurry and residue, BLAS poses much less environmental pressure on the surrounding environment. However, it has the shortcoming of inefficient utilization of local resources and little contribution to local economy. Thus, how to improve the production efficiency is crucial for the promotion of BLAS. In addition, the system diversity ratio and dominance degree index are relatively low due to the imbalanced production structure of the whole system. Hence, to increase the breeding and biogas yield is another important work to strengthen the diversification of the BLAS.

As to the sustainability, the inner metabolism of energy and material flow of the BLAS (reuse of biogas slurry and residue) would alleviate environmental stress and achieve sustainable breeding and planting system.

7.2 Scenarios Analysis Based on Emergy Analysis

We provide a concise overview of the current status of a typical BLAS and explore future options for the system development by combining emergy synthesis with emergetic ternary diagram. Moreover, two major methods employed, emergy analysis and emergetic ternary diagrams, are introduced, and the case study in Gongcheng County and the future developmental scenarios are described. Furthermore, system performance of BLAS and its subsystems are addressed based on emergy evaluation results and emergetic ternary diagrams.

7.2.1 Materials and Methods

(1) **Emergy synthesis (EmS)**

Figure 7.4 shows the aggregated system emergy flows of a typical BLAS. The emergy inputs are parsed into three categories (Giannetti et al. 2006): renewable resources (R), nonrenewable resources (N), and the economic inputs (F). The R and N flows are economically free input offered by the local environment, of which the renewable resources can recover at least at the same rate as they are consumed, while the nonrenewable resources are depleted without recuperation. The F flows are provided by the market and are measured in currency. The emergy yield (Y) represents the output of total emergy produced or released by the BLAS.

Emergy-based indicators offer effective sustainability metrics to evaluate the utilization of bioenergy and biomass and track the ecological and economic behaviors of BLAS. Five basic emergy-based indices are as below, which are established for the comprehensive evaluation of BLAS with respect to system

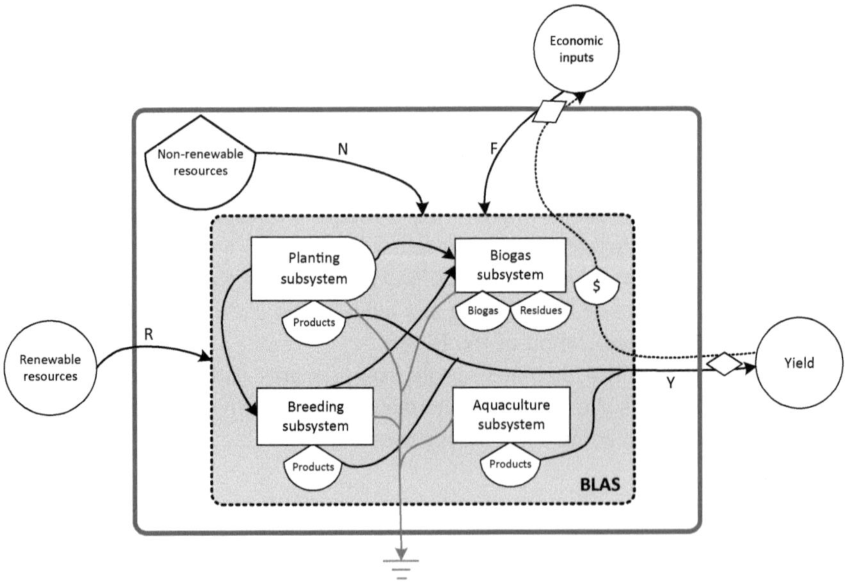

Fig. 7.4 Aggregated emergy flow diagram for BLAS. Reprint from Chen and Chen (2012), with permission from Elsevier

efficiency, yield, environmental load, and sustainability level. More details of the significations of these indices can be found in Campbell (1998) and Brown and Ulgiati (2004a, b).

(2) **Emergy ternary diagram (EmTD)**

Emergy ternary diagram (EmTD) is an integrated tool that combines emergy accounting with ternary diagram (Giannetti et al. 2006; Almeida et al. 2007). Different from the other emergy representation forms (e.g., Tonon et al. 2000; Giannantoni et al. 2002), EmTD functioned as a diagram of exhibiting emergy evaluation results and a data treatment platform. The special data disposal and graphic representation pattern provided by EmTD make it possible to compare various processes and systems, evaluate, and track down changes of system performance over time. With the aid of ternary diagrams, aspects such as the interactions inside of one system and the interactions between systems and the environment can be recognized and evaluated.

EmTD evaluates the system based on three emergy categories: *R*, *N*, and *F*. Each apex in an equilateral triangle represented a category (the proportions of three categories determine a point inside of this triangle). The relative proportions of each element are given by the lengths of the perpendicular from the given point to the side opposite to the corresponding element of the triangle. In such a way as defined, ternary combinations that reflect the changes of the concerned emergy system are represented by points within the triangle. Besides, it can also visually represent some of the emergy indices (i.e., ELR, EYR, EIR, SI) which are formulated by *R*, *N,* and *F*. In that way, the comparison of different development paths or strategies can be achieved, which is critical for identifying the sustainability scenarios of BLAS.

In fact, besides the regular graphic representation, EmTD betters itself in the release or identification of some component-specific traits and system evolution trends, therefore investigating the system performance in a more explicit way compared to the traditional interpretations. For these, we concluded in the four aspects as follows:

System aggregation analysis (Fig. 7.5a): EmTD allows an aggregated analysis of various combinations of subsystems. When two different ternary compositions (subsystems), say point *p* and *q* within the triangle, are jointly considered, the combined system can be represented by a resultant point termed as simergic point (*S*) that lies at somewhere between *p* and *q*.

Input sensitivity analysis (Fig. 7.5b): Based on input sensitive lines, it is possible to track system's developmental trends. The changes of the lengths of the perpendicular from the given point (along the straight line) to the side opposite to the corresponding element of the triangle would represent the quantity changes of the corresponding element of the apex. Besides, any point along this line shares the

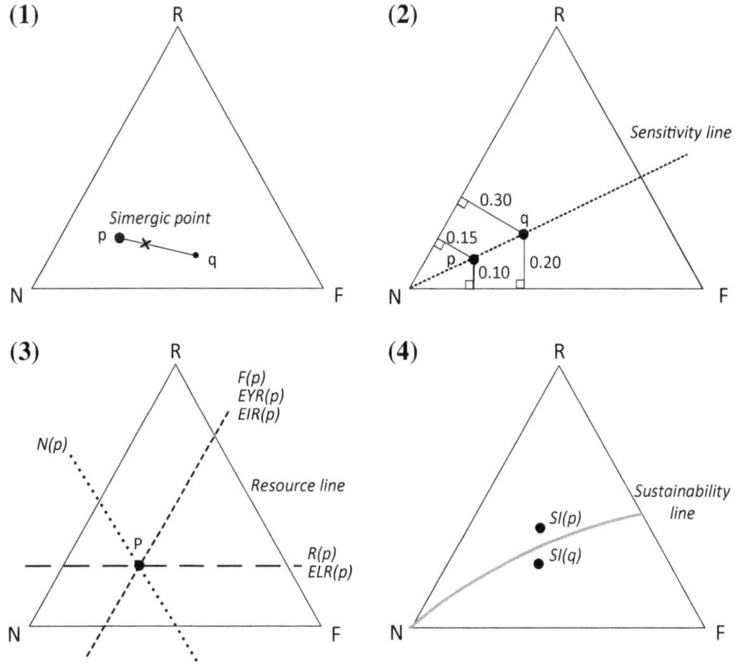

Fig. 7.5 Emergetic ternary diagrams for *1* system aggregation analysis, *2* input sensitivity analysis, *3* resource equivalence analysis, and *4* sustainability zone analysis. Reprinted from Chen and Chen (2012), with permission from Elsevier

same initial proportion of other two elements represented by other two apexes. For example, the system illustrated below is progressively poorer in *N*, as it passes from "*p*" to "*q*," leaving *R* and *F* at the same initial proportion.

Resource equivalence analysis (Fig. 7.5c): For example, *R* is represented by the relative resource line (*R* resource line) that is parallel to the triangle side opposite to the *R* element apex within the diagram. It is very useful for comparing the usage of resources by-products or processes. As the sum of *R*, *N*, and *F* equals 100% by definition, three emergy-based indices can be further formulated as follows: EYR = $100/F$, EIR = $F/(100 - F)$ (shown with the *F* resource line), and ELR = $(100 - R)/R$ (shown with the *R* resource line).

Sustainability zone analysis (Fig. 7.5d): EmTD provides a convenient way to indicate constant values of the sustainability index. In the ternary diagram, SI can be further formulated as $100R/(100 - R)\ F$. The corresponding sustainability lines turn out to be specific curves that depart from the N apex in the direction of the RF side, allowing divisions of the triangle into distinct sustainability zones (e.g., points *p* and *q* are in different sustainability zones that departed by a sustainability line, mostly according to the criteria depicted in Table 7.9), which are indicative when identifying and comparing the sustainability of different systems and processes.

Table 7.9 Emergy-based indices for BLAS evaluation (Brown and Ulgiati 2004a, b; Odum 1996; Ulgiati and Brown 1998)

Symbol	Formulation	Implication
BEE	Y_b/F	i.e., biogas energy efficiency, it is a metric of the production efficiency of biogas energy, calculating by the ratio of produced biogas (Y_b) to economic input (F)
EMI	EM/U	i.e., emission mitigation intensity, it measures the net emission mitigation (EM) for per unit emergy input
EYR	U/F = (F + R + N)/F	i.e., environmental yield ratio, it is a measure of the ability of a process to exploit available locally renewable and nonrenewable resources by investing outsider sources. The higher the value of this index, the greater the return obtained per unit of emergy invested
ELR	(N + F)/R	i.e., environmental loading ratio, it is a measure of the possible disturbance to the local drive from outside sources. The lower the ratio is, the lower the stress is to the environment When 3 < ELR < 10, the impact is considered moderate; when ELR > 10, there is a high environmental loading; extremely high values of ELR: The offer of local renewable inputs is not enough to supply the process demands
SI	EYR/ELR	i.e., sustainability index, it is an aggregating index based on both interaction with the surrounding environment and renewability. When SI < 1, products and processes are not sustainable in a long term; 1 < SI < 5: point E, may make a sustainable contribution to the economy for medium periods and processes; SI > 5: sustainable in a long term

(3) **Description of future scenarios**

The development of BLAS is being stimulated by both existing and forthcoming governmental incentives. Three scenarios for 2015 are hypothetically designed based on an explicit policy investigation of biogas energy and eco-agricultural development at local as well as country scale, while another two scenarios are set by a combination of the former ones. In order to explore the future developmental direction and sustainability of BLAS, these scenarios are mainly focused on aspects that would potentially contribute to the changes in emergy input, structure of BLAS, and its subsystems (such as biogas construction, investment expanding, and technical renovation). Descriptions of scenario settings and corresponding changes with respect to changes of system input are listed in Table 7.10 (the items of the corresponding changes are within the emergy input list in Table 7.11).

Table 7.10 Descriptions of five developmental scenarios of BLAS

Scenario	Description
Scenario I (S1)	※ *Scenario setting* Another 6500 biogas digesters are built before 2015 (i.e., approximately 10% of the existing biogas digesters)
	※ *Responsive changes* Soil loss (−20%), labor and management (−10%), organic fertilizer (−30%), electricity (−10%), biogas construction and maintenance (+10%), nitrogen (−20%), phosphate (−20%), potash (−20%), compound fertilizer (−20%), pesticide (+20%), infrastructures (+10%), ceteris paribus
Scenario II (S2)	※ *Scenario setting* The investments in planting, breeding, and aquaculture industry are all increased by 30%, while the number of biogas digesters maintains status quo
	※ *Responsive changes* Soil loss (+30%),labor and management (+20%), seedlings (+30%), grains (+30%), organic fertilizer (+20%), electricity (+10%), diesels (+30%), machinery (+20%), pig feed (+30%), fish feed (+30%), plastic films (+20%), nitrogen (+20%), phosphate (+20%), potash (+20%), compound fertilizer (+20%), pesticide (+20%), infrastructures (+30%), ceteris paribus
Scenario III (S3)	※*Scenario setting* Chemical fertilizer and pesticide inputs are cut by 30% with an enhancement of the comprehensive use of biogas slurry and residue, and also, the construction cost of biogas facilities is reduced by 40% due to technical renovation, while the number of Biogas digesters maintain the status quo
	※ *Responsive changes* Labor and management (−20%), organic fertilizer (−30%), electricity (−10%), nitrogen (−30%), phosphate (−30%), potash (−30%), compound fertilizer (−30%), pesticide (−30%), infrastructures (−30%), ceteris paribus
Scenario IV (S12)	※ *Scenario setting* A combination of Scenario I and Scenario II
Scenario V (S13)	※ Scenario setting A combination of Scenario I and Scenario III

Table 7.11 Emergy evaluation of overall BLAS

Number	Item	Raw data	Unit	Transformity[h]/ (sej/unit)	References	Solar emergy/ (sej)
1	Solar radiation[a]	9.90E+18	J	1.00E+00	By definition	9.90E+18
2	Rain, chemical[b]	1.53E+16	J	3.06E+04	Odum (1996)	4.68E+20
3	Rain, geopotential[c]	2.43E+16	J	1.49E+04	Odum (1996)	3.63E+20
4	Wind, kinetic[d]	1.49E+17	J	1.11E+03	Odum (1996)	1.66E+20
5	Earth cycle[e]	2.15E+15	J	4.87E+04	Odum (1996)	1.05E+20
Total *R* (only 2, 5 areaccounted)						5.73E+20
6	Soil loss[f]	9.65E+14	J	1.24E+05	Brown and Arding (1991)	1.20E+20
Total *N*						1.20E+20
7	Labor and services[g] (L&S)	1.79E+07	US $	5.87E+12	Yang et al. (2010)	1.05E+20
8	Seedlings	1.97E+05	US $	5.87E+12	Yang et al. (2010)	1.16E+18
9	Grains	5.66E+13	J	1.39E+05	Lan et al. (2002)	7.89E+18
10	Organic fertilizer	1.71E+13	J	4.54E+06	Odum (1996)	7.76E+19
11	Electricity	7.74E+12	J	3.36E+05	Odum (1996)	2.60E+18
12	Diesels	2.94E+13	J	1.11E+05	Odum (1996)	3.26E+18
13	Machinery	2.50E+09	g	1.12E+10	Odum (1996)	2.80E+19
14	Biogas construction and maintenance[h]	7.80E+05	US $	5.87E+12	Yang et al. (2010)	4.58E+18
15	Pig feed	7.12E+06	US $	5.87E+12	Yang et al. (2010)	4.18E+19
16	Fish feed	4.05E+05	US $	5.87E+12	Yang et al. (2010)	2.38E+18
17	Plastic films	5.88E+09	g	1.11E+05		6.53E+14
18	Nitrogen	2.55E+10	g	6.38E+09	Odum (1996)	1.63E+20
19	Phosphate	1.15E+10	g	6.55E+09	Odum (1996)	7.53E+19
20	Potash	9.03E+09	g	1.74E+09	Brandt-Williams (2002)	1.57E+19
21	Compound fertilizer	1.80E+10	g	4.70E+09	Lan et al. (2002)	8.47E+19
22	Pesticides	2.38E+09	g	2.52E+10	Brown and Arding (1991)	6.00E+19
23	Infrastructure	3.01E+05	US $	5.87E+12	Yang et al. (2010)	1.77E+18
Total *F*						6.74E+20
24	Planting products	2.54E+15	J	8.30E+04	Lan et al. (2002)	2.11E+20
25	Breeding products	5.38E+14	J	1.71E+06	Lan et al. (2002)	9.20E+20
26	Aquaculture products	2.50E+13	J	2.00E+6	Lan et al. (2002)	5.00E+19
27	Biogas	5.35E+14	J	2.48E+05	Bastianoni et al. (2001)	1.33E+20

(continued)

Table 7.11 (continued)

Number	Item	Raw data	Unit	Transformity[h]/ (sej/unit)	References	Solar emergy/ (sej)
Yield						1.31E+21
EM[i]		1.21E+11	g			
Total input with L&S		1.37E+21	sej			
Total input without L&S		1.26E+21	sej			
Empower density[j]		6.37E+11	sej/m^2			
Unit empower[k]		5.78E+15	sej			
Transformity with L&S		3.77E+05	sej/J			
Transformity without L&S		3.46E+05	sej/J			
Specific emergy with L&S[l]		1.53E+10	sej/g			
Specific emergy without L&S		1.40E+10	sej/g			

[a]Solar energy = (The average radiation of Guangxi Province) × (Area) = (4.61E+9 J/m^2/year) × (2.15E+9 m^2) = 9.90E+18 J/year; Average radiation in Guangxi Province is 4.61E+9 J/m^2/year

[b]Rain energy (chemical) = (Area) × (Average rainfall) × (Density) × (Gibbs free energy) = (2.15E+9 m^2) × (1.44 m/year) × (1.00E+6 g/m^3) × (4.94 J/g) = 1.53E+16 J/year; Average rainfall in Guangxi Province is 1.44 m/year (Cheng 1990)

[c]Rain energy (geopotential) = (Area) × (Mean altitude) × (Average rainfall) × (Density) ×(Acceleration of gravity) = (2.15E+9 m^2) × (800 m) × (1.44 m/year) × (1.00E+10^3 kg/m^3) ×(9.8 m/s^2) = 2.43E+16 J/year (Cheng 1990)

[d]Wind energy = (Area) × (Air density) × (eddy diffusivity) × (Gradient Velocity) = (2.15E+9 m^2) × (1.30 kg/m^3) × (1.80 m^2/s)(3.15E+7 s/year) = 1.49E+17 J/year

[e]Energy of earth cycle = (Area)(Heat flux) = (2.15E+9 m^2)(1.00E+6 J/m^2/year) = 2.15E+15 J/year

[f]Energy of soil loss = (Area of agricultural land)(Soil loss rate)(organic matter content) × (5.40E+6 kcal/t) × (4186 J/kcal) = (1.82E+8 m^2) × (6.5E-03t/m^2/year) × (0.036) × (5.40E+6 kcal/t) × (4186 J/kcal) = 1.41E+15 J/year; Soil loss rate of Guangxi Province is 6.5E-03t/m^2/year, and organic matter content is 0.036 (He et al. 2009)

[g]Labor and services = (Area of agricultural land) × (Average wage) × (the exchange rate between RMB and US$ in 2000) = (1.82E+8 m^2) × (0.79RMB/m^2/year) × (1/8.3) = 1.79E+7US$/year

[h]All the transformities prior to the year 2000 are updated values, which were multiplied by 1.68

[i]The determination of the emission mitigation of BLAS is based on the assumption that it equals the scaling up of the emission mitigation effect of a typical biogas plant used in Gongcheng County, i.e., 8 m^3 household biogas plant (Wang et al. 2010). CO_2 and CH_4 emissions are considered here, i.e., EM = EM(CO_2) + EM(CH_4); EM(BLAS) = EM(CO_2) + EM(CH_4) = (1.41E+6 g + 4.93E+5 g) × 63600 = 1.21E+11 g

[j]Empower density = U/Area

[k]Unit empower = U/Population; the population of BLAS in Gongcheng County is 237,000

[l]Transformity = U/Total energy of system yield

[m]Specific emergy = U/Total biomass of system yield

7.2.2 Emergy-Based Analysis of BLAS

(1) Emergy synthesis

Emergy flows of overall BLAS are summed up in Table 7.11. The results show that the total emergy input of BLAS is 1.37E+21 sej (1.26E+21 sej without L&S). The energy of rain (chemical and geopotential) contributes most to the renewable energy of BLAS, while the labor and services (L&S) and chemical fertilizers (nitrogen, phosphate, potash, and compound fertilizer) serve as the two greatest emergy inputs, followed by organic fertilizer and pig feed. Based on the emergy flow evaluation, indices (such as empower density, unit empower, transformity, and specific emergy) can address energy flow intensity and system efficiency of BLAS is also evaluated. In addition, the aggregated emission mitigation magnitude of BLAS as opposed to traditional agricultural system is obvious.

The components of BLAS including emergy input and yield of the four subsystems (i.e., planting, breeding, aquaculture and biogas subsystem) are accounted and compared in Table 7.12. The results indicate planting subsystem is the biggest constituent of BLAS (U = 1.17E+21 sej), while the rest subsystems have similar performance. Thus, planting subsystem also owns the highest empower density and unit empower. Aquaculture and biogas subsystem have the highest and lowest transformities, respectively, either considering labor and service or not. Finally, EM results show the planting (EM = 7.93E+10 g) and breeding (EM = 2.79E+10 g) are the major contributors to emission mitigation in this system.

(2) **Emergetic ternary diagrams**

As demonstrated in Table 7.13, the overall BLAS is characterized by 42% renewable resources (R), 9% nonrenewable resources (N), and 49% economic input (F). The ratios of emergy input categories (R, N, and F) within planting and breeding subsystems are similar, while aquaculture and biogas subsystems need heavier economic inputs, respectively. The emergy structures of these systems are illustrated more intuitively in Fig. 7.6 in the form of emergetic ternary diagrams (EmTDs). Biogas energy efficiency (BEE) of BLAS is 0.20, implying that 0.20 biogas energy is produced when input 1.00 unit of energy as the same quality, while the BEE of biogas subsystem is relatively high (2.71). As to emission mitigation intensity (EMI), planting subsystem is the biggest contributor (0.57 g/10^{10} sej) followed by breeding, while biogas subsystem has a poor performance in mitigation.

7.2.3 *Scenario Analysis*

According to the five policy scenarios (S1, S2, S3, S12, and S13) elucidated previously, we evaluate the future alternatives of BLAS (Table 7.14). The N, R, F of each system (subsystem) within different scenarios are altogether accounted, and the relative emergy-based indices are comprehensively analyzed in contrast to the current status. Emergetic ternary diagrams of BLAS and its subsystems under these scenarios are also provided to demonstrate the holistic developmental trends and sustainability of BLAS and its subsystems, where the four analyses brought from EmTD technique are incorporated.

Table 7.12 Emergy evaluation of BLAS's subsystems

Number	Item	Solar emergy/(sej)			
		Planting	Breeding	Aquaculture	Biogas
1	Solar radiation	9.10E+18	2.00E+17	4.34E+17	1.66E+17
2	Rain, chemical	4.30E+20	9.45E+18	2.05E+19	7.86E+18
3	Rain, geopotential	3.33E+20	7.33E+18	1.59E+19	6.10E+18
4	Wind, kinetic	1.53E+20	3.35E+18	7.27E+18	2.79E+18
5	Earth cycle	9.63E+19	2.12E+18	4.59E+18	1.76E+18
Total R		5.26E+20	1.28E+19	2.78E+19	1.06E+19
6	Soil loss	1.10E+20	2.42E+18	5.24E+18	2.01E+18
Total N		1.10E+20	2.42E+18	5.24E+18	2.01E+18
7	Labor and services (L&S)	5.28E+19	1.39E+19	1.51E+19	2.32E+19
8	Seedlings	1.16E+18	–	–	–
9	Grains	7.89E+18	–	–	–
10	Organic fertilizer	5.82E+19	8.53E+18	1.09E+19	–
11	Electricity	7.80E+17	9.10E+17	6.50E+17	2.60E+17
12	Diesels	1.14E+18	8.15E+17	4.89E+17	8.15E+17
13	Machinery	1.26E+19	4.20E+18	4.20E+18	7.00E+18
14	Biogas construction and maintenance	–	–	–	4.58E+18
15	Pig feed	–	4.18E+19	–	–
16	Fish feed	–	–	2.38E+18	–
17	Plastic films	6.53E+14	–	–	–
18	Nitrogen	1.63E+20	–	–	–
19	Phosphate	7.53E+19	–	–	–
20	Potash	1.57E+19	–	–	–
21	Compound fertilizer	8.47E+19	–	–	–
22	Pesticides	6.00E+19	–	–	–
23	Infrastructure	8.83E+17	2.65E+17	1.77E+17	4.42E+17
Total F		5.34E+20	7.05E+19	3.38E+19	3.63E+19
24	Planting products	2.11E+20	–	–	–
25	Breeding products	–	9.20E+20	–	–
26	Aquaculture products	–	–	5.00E+19	–
27	Biogas	–	–	–	1.33E+20
Yield		2.11E+20	9.20E+20	5.00E+19	1.33E+20
EM[a]		7.93E+10	2.79E+10	0	1.36E+7
Total input with L&S		1.17E+21	6.68E+19	8.57E+19	4.90E+19
Total input without L&S		1.12E+21	5.29E+19	7.06E+19	2.58E+19
Empower density		5.44E+11	3.11E+10	3.99E+10	2.28E+10
Unit empower		4.94E+15	2.82E+14	3.62E+14	2.07E+14
Transformity with L&S		4.61E+05	1.24E+05	3.43E+06	9.16E+04

(continued)

Table 7.12 (continued)

Number	Item	Solar emergy/(sej)			
		Planting	Breeding	Aquaculture	Biogas
Transformity without L&S		4.40E+05	9.83E+04	2.82E+06	4.82E+04
Specific emergy with L&S		1.35E+09	2.85E+09	1.58E+10	2.70E+09
Specific emergy without L&S		1.29E+09	2.26E+09	1.30E+10	1.42E+09

[a]EM(Planting) = (1.19E+6 g + 5.45E+4 g) × 63600 = 7.93E+10 g; EM(Breeding) = (0 g + 4.38E+5 g) × 63600 = 2.79E+10 g; EM(Biogas) = (2.15E+5 g + 0 g) × 63600 = 1.36E+10 g; the emission mitigation of aquaculture can be neglected in the current system

Table 7.13 Emergy-based indices of BLAS and its subsystems

	R/(%)	N/(%)	F/(%)	BEE	EMI/(g/10^{10} sej)	EYR	EIR	ELR	SI
BLAS	43	9	48	0.20	0.88	2.03	0.97	1.39	1.46
Planting	45	9	46	–	0.57	2.19	0.84	1.22	1.79
Breeding	42	8	51	–	0.21	1.22	4.63	5.69	0.21
Aquaculture	15	3	82	–	0	1.97	1.03	1.41	1.40
Biogas	22	4	74	2.71	0.10	1.35	2.87	3.60	0.37

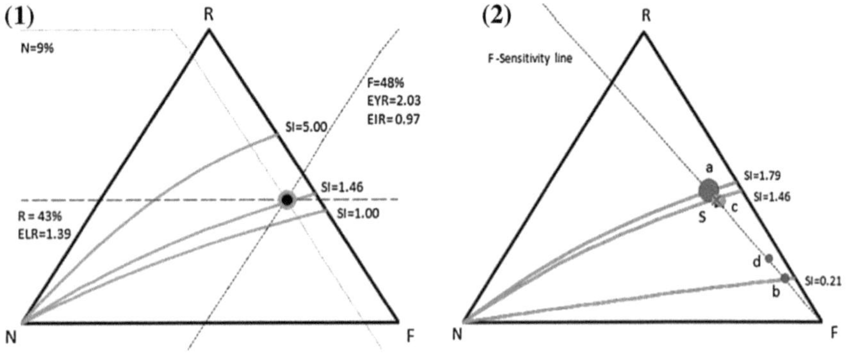

Fig. 7.6 Emergetic ternary diagrams for *1* BLAS and *2* its subsystems. *Notes Point a* planting subsystem; *Point b* breeding subsystem; *Point c* aquaculture subsystem; *Point d* biogas subsystem; *Point S* simergic point of all subsystems (also the BLAS's point calculating the overall system). *F*-sensitivity line is a straight line joining an apex (*F*) to point 1. Reprinted from Chen and Chen (2012), with permission from Elsevier

7.2.4 Discussion and Conclusions

Since the introduction of biogas energy to the agriculture industry chain, a huge economic development has taken place, which greatly improved the living standards of local resident. This can be observed from the higher empower density (6.37E+11sej/m^2) and unit empower (5.78E+15sej) in BLAS over the average level of traditional agriculture system in China (Chen et al. 2006) and Italy (Ulgiati et al. 1993). However, the relatively higher transformity (3.77E+05sej/J) of BLAS

Table 7.14 Aggregated emergy evaluation of BLAS under different scenarios

		R/(sej)(%)	*N*/(sej)(%)	*F*/(sej)(%)	EE[a]	EMI/(g/10^{10}sej)	YR	IR	LR	SI
S1	BLAS	5.73E+20(47)	9.57E+19(8)	5.61E+20(46)	0.26	1.08	2.19	0.84	1.15	1.91
	Planting	5.26E+20(50)	8.80E+19(8)	4.31E+20(41)	–	0.70	2.42	0.70	0.99	2.46
	Breeding	1.28E+19(16)	1.93E+18(2)	6.64E+19(82)	–	0.26	1.22	4.51	5.34	0.23
	Aquaculture	2.78E+19(46)	4.19E+18(7)	2.90E+19(48)	–	0	2.10	0.91	1.20	1.76
	Biogas	1.06E+19(23)	1.61E+18(3)	3.45E+19(74)	4.24	0.12	1.36	2.81	3.39	0.40
S2	BLAS	5.73E+20(37)	1.56E+20(10)	8.14E+20(53)	0.16	0.58	1.89	1.12	1.69	1.12
	Planting	5.26E+20(40)	1.43E+20(11)	6.42E+20(49)	–	0.36	2.04	0.96	1.49	1.37
	Breeding	1.28E+19(12)	3.14E+18(3)	8.87E+19(85)	–	0.13	1.18	5.57	7.18	0.16
	Aquaculture	2.78E+19(37)	6.81E+18(9)	4.09E+19(54)	–	0	1.85	1.18	1.72	1.08
	Biogas	1.06E+19(19)	2.62E+18(5)	4.28E+19(76)	3.11	0.09	1.31	3.22	4.26	0.31
S3	BLAS	5.73E+20(48)	1.20E+20(10)	5.08E+20(42)	0.26	1.33	2.36	0.73	1.10	2.15
	Planting	5.26E+20(51)	1.10E+20(11)	3.86E+20(38)	–	0.86	2.65	0.61	0.94	2.81
	Breeding	1.28E+19(16)	2.42E+18(3)	6.50E+19(81)	–	0.30	1.04	25.38	30.36	0.03
	Aquaculture	2.78E+19(46)	5.24E+18(9)	2.75E+19(45)	–	0	2.20	0.83	1.18	1.87
	Biogas	1.06E+19(25)	2.01E+18(5)	2.94E+19(70)	4.52	0.11	1.43	2.32	2.95	0.49
S12	BLAS	5.73E+20(41)	1.26E+20(9)	6.88E+20(50)	0.21	0.82	2.00	1.00	1.44	1.39
	Planting	5.26E+20(45)	1.15E+20(10)	5.37E+20(46)	–	0.53	2.20	0.84	1.24	1.77
	Breeding	1.28E+19(14)	2.54E+18(3)	7.76E+19(83)	–	0.20	1.20	5.06	6.26	0.19
	Aquaculture	2.78E+19(41)	5.50E+18(8)	3.49E+19(51)	–	0	1.95	1.05	1.46	1.34
	Biogas	1.06E+19(21)	2.12E+18(4)	3.86E+19(75)	3.79	0.11	1.33	3.03	3.83	0.35
S13	BLAS	5.73E+20(47)	1.08E+20(9)	5.35E+20(44)	0.27	1.21	2.28	0.79	1.12	2.03
	Planting	5.26E+20(51)	9.90E+19(10)	4.09E+20(40)	–	0.78	2.54	0.65	0.97	2.63
	Breeding	1.28E+19(16)	2.18E+18(3)	6.57E+19(81)	–	0.28	1.13	14.94	17.85	0.13
	Aquaculture	2.78E+19(46)	4.72E+18(8)	2.83E+19(47)	–	0	2.15	0.87	1.19	1.81
	Biogas	1.06E+19(24)	1.81E+18(4)	3.20E+19(72)	4.57	0.12	1.39	2.57	3.17	0.44

[a]The values of BEE and EMI in scenario analysis are estimated based on the changes of system input (see Table 2), e.g., BEE = (BLAS, S1) = Y_b/F = (1.33E+20 × 110%)/5.61E+20 = 0.26, EMI (BLAS, S1) = EM/U = (1.41E+6 g + 4.93E+5 g) × (63600 × 110%)/1.23E+21 = 1.08 g/10^{10} sej

compared to other agricultural energy systems (such as Wei et al. 2009; Dong et al. 2008; Lan et al. 2002; Zhang et al. 2007a, b) is a disadvantage. Emergetic ternary diagrams indicate another fact that is in contrast to the relatively stable renewable and nonrenewable resources. The maintenance of current BLAS is highly reliant on economic input (*F*). As unveiled by the *F*-sensitivity line, this developmental trait is actually shared by all the subsystems in BLAS. The overall BLAS, planting subsystem, and aquaculture subsystem are within 1–5 sustainability zone in all scenarios, indicating the sustainable development is possible for them in the medium term, while breeding and biogas remain in an unsustainable situation constrained in 0–1 sustainability zone, which is due to the high environmental load carried by breeding and biogas subsystem ($3 < \mathrm{ELR} < 10$). In view of the severe variation between two marginal F resource lines shown in EmTDs, the heavy proportion of economic input just mentioned also poses a big threat for whole system's sustainable development. In terms of the further development of BLAS, extra biogas constructions and effective technical innovation for biogas comprehensive utilization and cost reduction may prove better alternatives over simple expansion of planting and breeding investment.

Emergetic ternary diagrams of BLAS and its subsystems under five policy scenarios (S1, S2, S3, S12 and S13) are provided to demonstrate the holistic development trends and sustainability of BLAS and its subsystems, where the four analyses brought from EmTD technique are incorporated (Fig. 7.7). It should be noted that the conflict between emission mitigation and system growth (which always implies growth of energy utilization) is not found in BLAS. Instead, positive relations are discovered among carbon emissions mitigation, efficient utilization of biogas energy, and sustainable development of overall biogas-linked agro-ecosystem, as shown in the regressions between each two of the three pertinent emergy-based indicators (EMI, SI, and BEE), whose effect is most significant between emission mitigation intensity and biogas energy efficiency (Fig. 7.8). It is also clear that the subset constituted by S3, S13, and S1 served as the best-performance alternative for a better BLAS, while S2 remains to be the worst-performance option. With the expansion of demand of energy for feeding the rapid growth of China's economy, acquiring clean and safe energy sources will be put at the first position at the development agenda. The BLAS in China investigated that in this chapter, the connection of renewable energy to the local agriculture and industry shows a preferred pattern in sustainable development. Therefore, BLAS is a promising option to sustain rural energy transition and energy safety of the country in the face of the forthcoming energy scarcity.

7.3 Three-Level Emergetic Evaluation

A novel emergetic evaluation pattern (so-called three-level emergetic evaluation framework) was developed to explore the renewability and efficiency of the BLAS, in which the system-wide performance, production process traits, and resource supplying behaviors linked by household biogas construction were tracked and

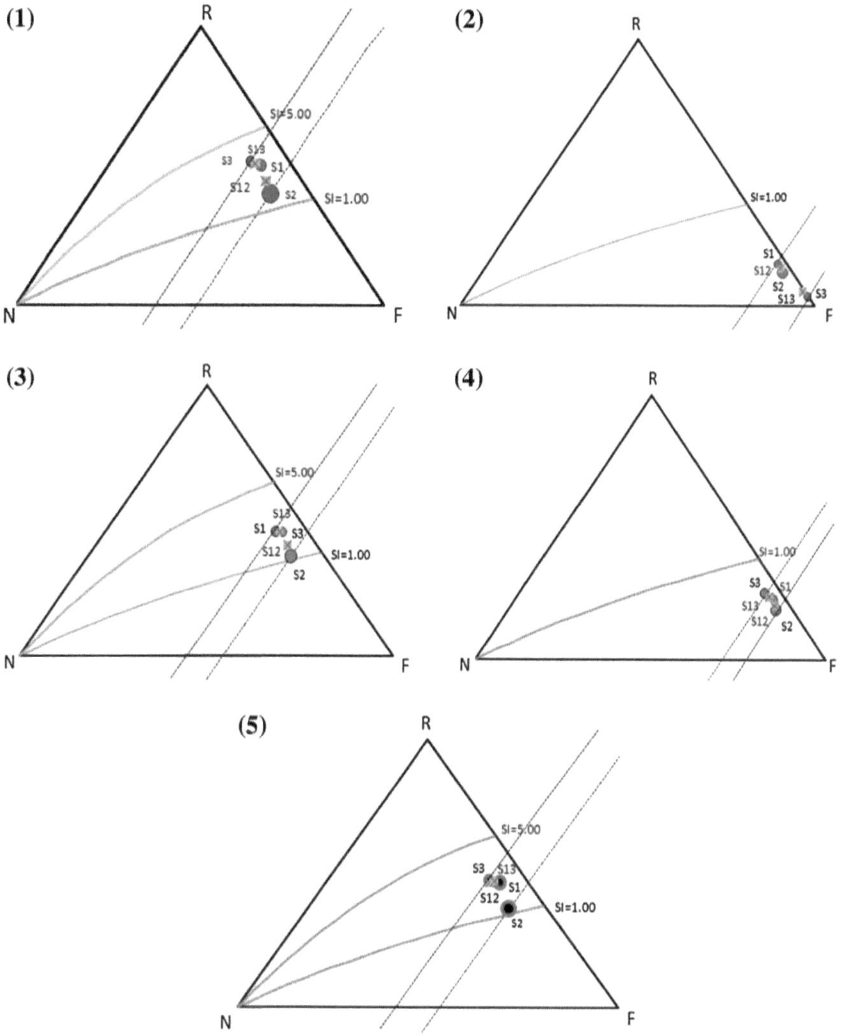

Fig. 7.7 Emergetic ternary diagrams of BLAS under five scenarios for *1* planting subsystem, *2* breeding subsystem, *3* aquaculture system, *4* biogas subsystem, and *5* overall BLAS. *Note* The size of system points is proportional to their total emergy input values. Reprinted from Chen and Chen (2012), with permission from Elsevier

appraised. By adapting the new framework to the existing emergetic analysis system, both the macro-properties of the complex system and the microdynamics underneath the system can be aggregately unveiled for potential improvement on concrete production activities. The aims of this study are as follows: (1) apprehend the renewability and efficiency of BLAS from a thermodynamical perspective for better management of the system at both macroscopic and microscopic scales; (2) compare the biophysical performances of different production processes of

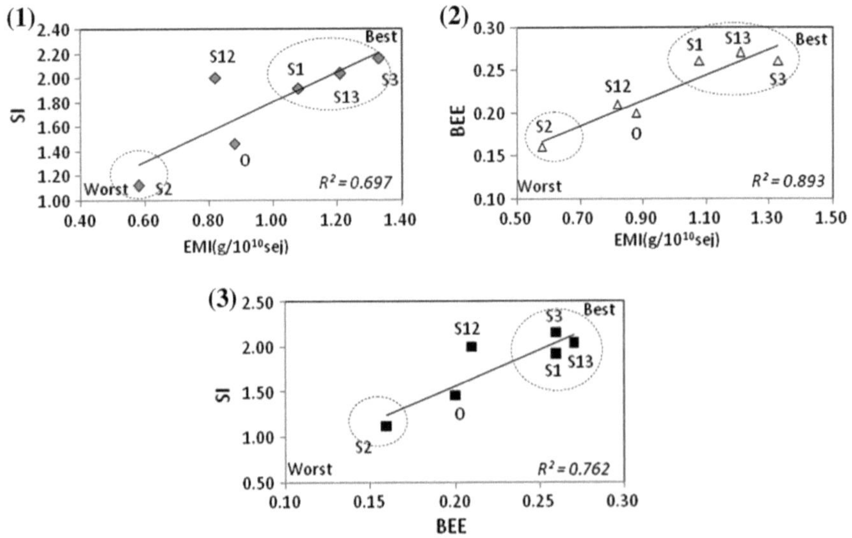

Fig. 7.8 Regression relations between EMI and SI (*1*), EMI and BEE (*2*), and BEE and SI (*3*) for current status (represented as "O") and five scenarios of BLAS. Reprinted from Chen and Chen (2012), with permission from Elsevier

BLAS and all the resource components within these processes over time; (3) derive an explicit analytical pattern (or framework) that permits the examination and optimization of an ecological–economic system at multiple levels.

7.3.1 Materials and Methods

(1) **Emergy metaphysics and accounting**

In terms of the accounting of emergy flows within a system, four categories of resource are identified generally, i.e., renewable resources (RR), nonrenewable resources (NR), renewable purchases (RP), and nonrenewable purchases (NP). These four elements are fundamental to emergy accounting and the understanding of the system interactions with the environment. The combination of them makes the total emergy input (U_{total}). The total of the resultant product or service outputs derived from these inputs is the emergy yield (Y_{total}).

(2) **Three-level emergetic evaluation**

A three-level evaluation framework based on emergy was developed, comprising three analytical fractions: overall system accounting (Level I), processed-based evaluation (Level II), and resource component tracking (Level III), to evaluate the renewability and efficiency of BLAS from different perspectives. The new

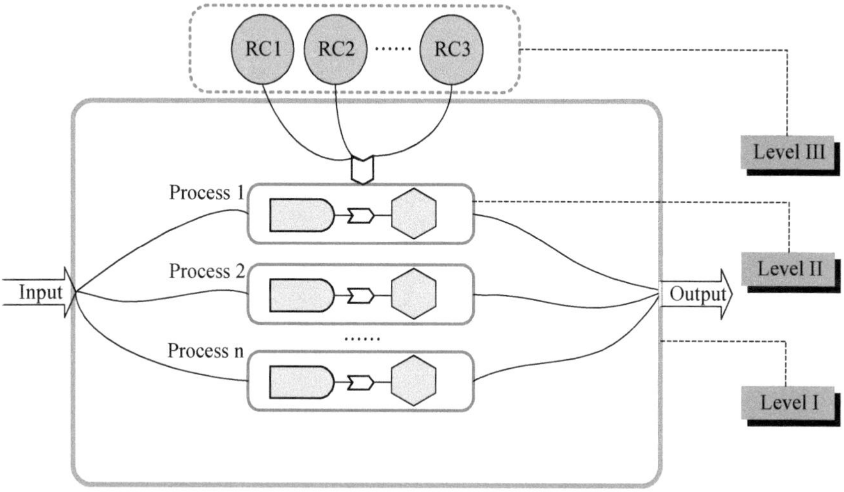

Fig. 7.9 Conceptual framework of three-level emergetic evaluation. *Note* RC stands for resource component. Reprinted from Chen and Chen (2014), with permission from Elsevier

evaluation framework is illustrated in Fig. 7.9. For the first level, key emergy indices include output efficiency (the ratio of total yield to total input), transformity (unit emergy value of a process), emergy yield ratio (EYR), environmental load ratio (ELR), emergy sustainability index (ESI).

Two new indices were developed, i.e., emergy contribution ratio (ECR) and emergy supply efficiency (ESE), based on the existing emergy index system. These two indices were primarily defined on the resource components' base to achieve level II evaluation and then applied them to the process-based evaluation by considering all components involved in this process to achieve level III evaluation. ECR is designed to reflect the direct and indirect contribution of a specific resource in a given process it is engaged in. Therefore, it is the ratio of the emergy input from a specific type of resource to the average emergy contribution of that process. When ECR > 1, it means the contribution of a specific resource to the production process it involved exceeds the average contribution level of the system, and vice versa. The resource with a high value of ECR plays a more significant role in the production process. ESE is the proportion of yield of a resource component to the average yield ratio regarding the whole system, implying the efficiency of a component in a given process. When ESE > 1, it means the supply efficiency of one resource in the relative production process exceeds the average level, and vice versa. A higher ESE indicates the given resource is used more intensively during the production process.

$$\mathbf{U} = (u_{i,j})_{\lambda \times \eta} \tag{7.6}$$

$$\mathbf{T} = (t_{i,1})_{\lambda\times 1}, t_{i,1} = \sum_{j=1}^{\eta} u_{i,j} \tag{7.7}$$

$$U_{\text{total}} = \sum_{i=1}^{\lambda} t_{i,1} \tag{7.8}$$

$$\mathbf{Y} = (\mathrm{y}_{i,1})_{\lambda\times 1} \tag{7.9}$$

$$Y_{\text{total}} = \sum_{i=1}^{\lambda} \mathrm{y}_{i,1} \tag{7.10}$$

$$\mathbf{ECR} = (c_{i,j})_{\lambda\times\eta}, c_{i,j} = \frac{u_{i,j}}{t_{i,1}/\hat{\eta}} = \frac{\hat{\eta}u_{\mathrm{i,j}}}{t_{i,1}} \tag{7.11}$$

$$\mathbf{ESE} = (\mathrm{s}_{i,j})_{\lambda\times\eta}, \mathrm{e}_{i,j} = \frac{y_{i,1}/u_{i,j}}{\eta Y_{\text{total}}/U_{\text{total}}} = \frac{y_{i,1}U_{\text{total}}}{\eta u_{i,j}Y_{\text{total}}} \tag{7.12}$$

where $u_{i,j}$ is the emergy input of ith resource component for jth production process; $t_{i,1}$ is the total emergy input of 1th process; $y_{i,1}$ is the emergy yield of 1th process; $c_{i,j}$ and $\mathrm{s}_{i,j}$ denote the ECR and ESE values of each combination of component and process; $\hat{\eta}$ is the number of components that actually contribute to the processes (i.e., $u_{i,j} > 0$), which is different from η (all components involved in the whole

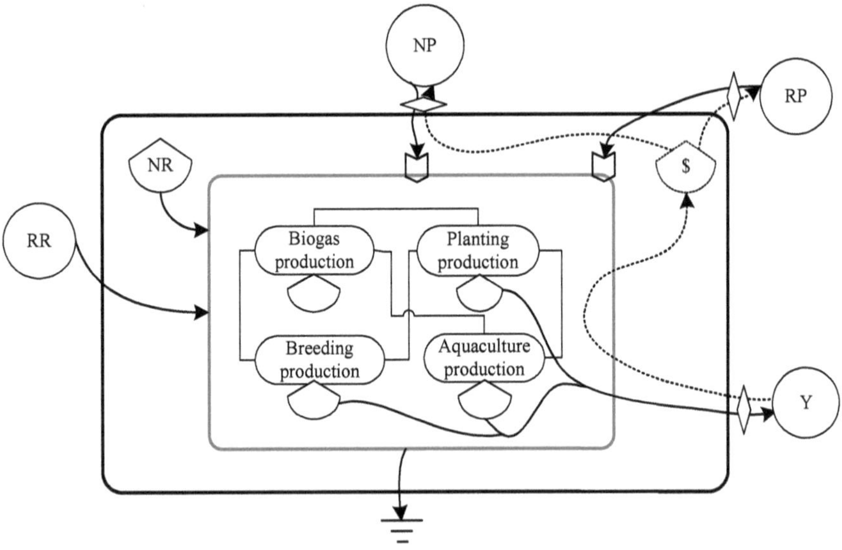

Fig. 7.10 Aggregated emergy flow diagram of a BLAS. *Notes RR* renewable resources; *NR* nonrenewable resources; *NP* nonrenewable purchases; *NP* nonrenewable purchases. Reprinted from Chen and Chen (2014), with permission from Elsevier

system). $u_{i,j}$, $t_{i,1}$, $y_{i,1}$, $c_{i,j}$, and $s_{i,j}$ are treated as elements of matrixes $\mathbf{U}_{\lambda\times\eta}$, $\mathbf{T}_{\lambda\times1}$, $\mathbf{Y}_{\lambda\times1}$, $\mathbf{ECR}_{\lambda\times\eta}$, and $\mathbf{ESE}_{\lambda\times\eta}$, respectively. Table 7.15 provided the pattern to calculate ESE and ECR.

(3) **System diagram**

7.3.2 *Results and Discussion*

(1) **Overall system accounting (Level I)**

The emergy flows of BLAS in Gongcheng County in years 2000, 2004, and 2008 were accounted, respectively. The boundary and structure of the system were depicted in Fig. 7.10. Table 7.16, Table 7.17 and Table 7.18 show the results for year 2008, 2004 and 2000, respectively. For a better comparison, all the evaluations were based on the per capita level.

The results showed that the BLAS has grown bigger over the years (emergy input from 4.22E+15 sej in 2000 to 4.64E+15 sej in 2008; emergy yield from 3.14E+15 sej in 2000 to 4.17E+15 sej in 2008) and displays an accelerating tendency. The structure of system emergy input was relatively stable: the RR (e.g., 42.5% in 2008) contributed most toward the total emergy input, followed by NP (34.9%), RP (14.1%), and NR (8.4%). The soil loss of the agro-system has been reduced (from 4.15E+14 sej to 3.88E+14 sej), because of the exploitation of agricultural land. Although the economic input (NP and RP) has increased over time (from 1.83E+15 sej to 2.28E+15 sej), a trend toward a more renewable condition of the BLAS has been unveiled with the increasing of RP's proportion (from 4.8 to 14.1%) and the decreasing of NP's (from 38.6 to 34.9%). The economic input includes chemical fertilizers (including nitrogen, compound fertilizer, phosphate, and potash), pesticides, labor and management, and organic fertilizers. Specifically, chemical fertilizers still contributed more than half of the total economic input in 2008. Organic fertilizers have been more frequently used due to the expansion of biogas plants, with an increase from 2.5% in 2000 to 11.7% in 2008. Also, the emergy input of labor and management has increased due to the increased job opportunities offered by the motivation of biogas to agricultural industries. Other purchases, electricity, diesels, machinery, pig feed, and fish feed, have been escalated too to accompany with the biogas utilization. As to emergy yield, all the agricultural products have witnessed more or less increment: meat product (from 2.32E+15 sej to 3.14E+15

Table 7.15 Analytical table for emergetic evaluation at levels II and III

Process\component	1	2	3	…	j	…	η
1							
2							
…							
i							
…							
λ							

Table 7.16 Emergy accounting of a BLAS in Gongcheng County (2008)

No.	Item	Raw data	Unit	Transformity/ (sej/unit)	References	Solar emergy/ (sej)
1	Solar radiation[a]	3.41E+13	J	1.00E+00	By definition	3.41E+13
2	Rain, chemical[b]	5.28E+10	J	3.06E+04	Odum (1996)	1.61E+15
3	Rain, geopotential[c]	8.38E+10	J	1.49E+04	Odum (1996)	1.25E+15
4	Wind, kinetic[d]	5.14E+11	J	1.11E+03	Odum (1996)	5.72E+14
5	Earth cycle[e]	7.41E+09	J	4.87E+04	Odum (1996)	3.61E+14
Total renewable resources (TRR)						1.97E+15
6	Soil loss[f]	3.13E+09	J	1.24E+05	Brown and Arding (1991)	3.88E+14
Total nonrenewable resources (TNR)						3.88E+14
7	Labor and management[g]	6.07E+01	US $	5.87E+12	Yang et al. (2010)	3.56E+14
8	Seedlings	6.89E-01	US $	5.87E+12	Yang et al. (2010)	4.05E+12
9	Grains	1.97E+08	J	1.39E+05	Lan et al. (2002)	2.75E+13
10	Organic fertilizer	5.90E+07	J	4.54E+06	Odum (1996)	2.67E+14
Total renewable purchases (TRP)						6.55E+14
11	Electricity	2.87E+07	J	3.36E+05	Odum (1996)	9.64E+12
12	Diesels	5.48E+07	J	1.11E+05	Odum (1996)	6.08E+12
13	Machinery	8.62E+03	g	1.12E+10	Odum (1996)	9.66E+13
14	Biogas construction and maintenance[h]	4.08E+00	US $	5.87E+12	Yang et al. (2010)	2.39E+13
15	Pig feed	2.11E+01	US $	5.87E+12	Yang et al. (2010)	1.24E+14
16	Fish feed	1.10E+00	g	5.87E+12	Bastianoni et al. (2001)	6.43E+12
17	Nitrogen	8.70E+04	g	6.38E+09	Odum (1996)	5.55E+14
18	Phosphate	3.88E+04	g	6.55E+09	Odum (1996)	2.54E+14
19	Potash	3.10E+04	g	1.74E+09	Brandt-Williams (2002)	5.40E+13
20	Compound fertilizer	5.96E+04	g	4.70E+09	Lan et al. (2002)	2.80E+14
21	Pesticides	8.22E+03	g	2.52E+10	Brown and Arding (1991)	2.07E+14
22	Infrastructure	1.09E+00	US $	5.87E+12	Yang et al. (2010)	6.39E+12
Total nonrenewable purchases (TNP)						1.62E+15
23	Crops	3.47E+09	J	8.30E+04	Odum (1996)	2.88E+14
24	Fruits	4.10E+09	J	2.70E+04	Odum (1996)	1.11E+14
25	Vegetables	6.10E+08	J	5.30E+04	Odum (1996)	3.23E+13
26	Teas	2.56E+07	J	2.00E+04	Odum (1996)	5.12E+11
27	Meat product (pig, cow, etc.)	1.83E+09	J	1.71E+06	Odum (1996)	3.14E+15

(continued)

Table 7.16 (continued)

No.	Item	Raw data	Unit	Transformity/ (sej/unit)	References	Solar emergy/ (sej)
28	Aquatic product	7.88E+07	J	2.00E+06	Odum (1996)	1.58E+14
29	Biogas	1.79E+09	J	2.48E+05	Bastianoni and Marchettini (2000)	4.45E+14
Total yield (TY)						4.17E+15

[a]Solar energy = (The average radiation of Gongcheng County) × (Area)/Population = (4.61E+9 J/m^2/year) × (2.15E+9 m^2)/2.90E+5 = 3.41E+13 J/year

[b]Rain energy (chemical) = (Area) × (Average rainfall) × (Density) × (Gibbs free energy)/Population = (2.15E+9 m^2) × (1.44 m/year) × (1.00E+6 g/m^3) × (4.94 J/g)/2.90E+5 = 5.28E+10 J/year; Average rainfall in Guangxi Province is 1.44 m/year (Cheng 1990)

[c]Rain energy (geopotential) = (Area) × (Mean altitude) × (Average rainfall) × (Density) × (Acceleration of gravity)/Population = (2.15E+9 m^2) × (800 m) × (1.44 m/year) × (1.00E+10^3 kg/m^3) × (9.8 m/s^2)/2.90E+5=8.38E+10 J/year (Cheng 1990)

[d]Wind energy = (Area) × (Air density) × (eddy diffusivity) × (Gradient Velocity)/Population = (2.15E+9 m^2) × (1.30 kg/m^3) × (1.80 m^2/s)(3.15E+7 s/year)/2.90E+5 = 5.14E+11 J/year

[e]Energy of earth cycle = (Area)(Heat flux)/Population = (2.15E+9 m^2)(1.00E+6 J/m^2/year)/2.90E+5 = 7.41E+9 J/year

[f]Energy of soil loss = (Area of agricultural land)(Soil loss rate)(organic matter content) × (5.40E+6 kcal/t) × (4186 J/kcal)/Population = (1.82E+8 m^2) × (6.13E-03t/m^2/year) × (0.036) × (5.40E+6 kcal/t) × (4186 J/kcal)/2.90E+5=3.13E+9 J/year; area of agricultural land of Gongcheng County in 2008 is 1.82E+8 m^2; soil loss rate of Guangxi Province is 6.13E-3t/m^2/year averagely, and organic matter content is 3.60% (He et al. 2009)

[g]Labor and services = (Area of agricultural land) × (Average wage) × (the exchange rate between RMB and US$ in 2000)/Population = (1.82E+8 m^2) × (0.80RMB/m^2/year) × (1/8.3)/2.90E+5=6.07E+1US$/year

[h]Money on biogas construction and maintenance: [(Number of new digester) × (construction fee) + (Number of digester underuse)(maintenance Production processes evaluation (Level II) fee)]/Population = [(3190) × (1650RMB/8.3/year) + (65100) × (70RMB/8.3/year)]/2.90E+5 = 4.08E+0US$/year

[i]All the transformities prior to the year 2000 are updated values, which were multiplied by 1.68

sej), biogas (from 3.16E+14 sej to 4.45E+14 sej), and crops (from 2.37E+14 sej to 2.88E+14 sej). Stimulated by biogas projects, the BLAS has been under a transaction process toward a higher renewability, while traditional agriculture still plays an indispensable part at this stage.

ii. **Emergetic evaluation (Level II)**

Based on the accounting results at the first level, the emergy flows within different production processes of the BLAS were further articulated and quantified. The results of years 2000, 2004, and 2008 were exhibited in Tables 7.19, 7.20, and 7.21, respectively. For the Level II emergetic evaluation, the overall BLAS was decomposed into four relatively independent production processes: planting, aquaculture, breeding, and biogas. All the empower values of the BLAS were divided and attached to the pertinent processes in which they were totally or partly utilized after examining the traits of resources supplied inside these processes. Emergy indices of different production processes within the BLAS were further

Table 7.17 Emergy accounting of the BLAS in Gongcheng County (2004)

No.	Item	Raw data	Unit	Transformity/ (sej/unit)	References	Solar emergy/ (sej)
1	Solar radiation	3.41E+13	J	1.00E+00	By definition	3.41E+13
2	Rain, chemical	5.67E+10	J	3.06E+04	Odum (1996)	1.73E+15
3	Rain, geopotential	9.00E+10	J	1.49E+04	Odum (1996)	1.34E+15
4	Wind, kinetic	5.15E+11	J	1.11E+03	Odum (1996)	5.73E+14
5	Earth cycle	7.41E+09	J	4.87E+04	Odum (1996)	3.61E+14
Total renewable resources (TRR)						2.10E+15
6	Soil loss	3.21E+09	J	1.24E+05	Brown and Arding (1991)	3.98E+14
Total nonrenewable resources (TNR)						3.98E+14
7	Labor & management	4.17E+01	US $	5.87E+12	Yang et al. (2010)	2.45E+14
8	Seedlings	3.66E-01	US $	5.87E+12	Yang et al. (2010)	2.15E+12
9	Grains	1.47E+08	J	1.39E+05	Lan et al. (2002)	2.05E+13
10	Organic fertilizer	2.10E+07	J	4.54E+06	Odum (1996)	9.51E+13
Total renewable purchases (TRP)						3.63E+14
11	Electricity	2.36E+07	J	3.36E+05	Odum (1996)	7.94E+12
12	Diesels	4.90E+07	J	1.11E+05	Odum (1996)	5.44E+12
13	Machinery	8.02E+03	g	1.12E+10	Odum (1996)	8.98E+13
14	Biogas construction and maintenance	3.54E+00	US $	5.87E+12	Yang et al. (2010)	2.08E+13
15	Pig feed	1.70E+01	US $	5.87E+12	Yang et al. (2010)	9.99E+13
16	Fish feed	8.96E-01	g	5.87E+12	Bastianoni et al. (2001)	5.26E+12
17	Nitrogen	8.67E+04	g	6.38E+09	Odum (1996)	5.53E+14
18	Phosphate	3.97E+04	g	6.55E+09	Odum (1996)	2.60E+14
19	Potash	3.34E+04	g	1.74E+09	Brandt-Williams (2002)	5.82E+13
20	Compound fertilizer	6.16E+04	g	4.70E+09	Lan et al. (2002)	2.90E+14
21	Pesticides	9.12E+03	g	2.52E+10	Brown and Arding (1991)	2.30E+14
22	Infrastructure	8.78E-01	US$	5.87E+12	Yang et al. (2010)	5.15E+12
Total nonrenewable purchases (TNP)						1.62E+15
23	Crops	3.05E+09	J	8.30E+04	Odum (1996)	2.53E+14
24	Fruits	4.07E+09	J	2.70E+04	Odum (1996)	1.10E+14
25	Vegetables	6.02E+08	J	5.30E+04	Odum (1996)	3.19E+13
26	Teas	2.35E+07	J	2.00E+04	Odum (1996)	4.70E+11

(continued)

Table 7.17 (continued)

No.	Item	Raw data	Unit	Transformity/ (sej/unit)	References	Solar emergy/ (sej)
27	Meat product (pig, cow, etc.)	1.54E+09	J	1.71E+06	Odum (1996)	2.64E+15
28	Aquatic product	6.78E+07	J	2.00E+06	Odum (1996)	1.36E+14
29	Biogas	1.45E+09	J	2.48E+05	Bastianoni and Marchettini (2000)	3.61E+14
Total yield (TY)						3.53E+15

calculated and compared on the basis of the decomposed accounting results (Table 7.22).

Taking year 2008 as an example (Table 7.21), the breeding is the most productive one with its highest total emergy yield. Biogas production has a higher output efficiency and contributes a lot to the overall efficiency. The output efficiency of the overall system was not simply equal to the combination of all the decomposed processes, which is interesting because they did not even share the same change trend after the decomposition. This is one reason why we need to dissect the overall system when we ponder over "what happens" at the process level. With respect to emergy transformation rate, the transformities of all the subprocesses took on the tendency of descent (either with or without the consideration of L&S), indicating the transformations of energy/resources have become more and more effective since the last decade. However, the problem of BLAS was revealed by its relatively high transformity (3.89E+05 sej/J in 2008) when compared with other agricultural systems (e.g., corn production in Italy (Ulgiati et al. 1993) and ethanol production from wheat in China (Hawking and Mlodinow 2010), mainly due to the relatively high national emergy money ratio of China, i.e., 5.87E+12 seJ/$ (only 1.93E+12 seJ/$ in U.S.) (Yang et al. 2010). By and large, the BLAS has made a medium-term sustainable contribution to the economy among all these years ($1 < ESI < 5$) [according to Ulgiati and Brown (2002)]. A challenge for this complex agro-system was unveiled by the continual descent of ESI within all the processes, resulting from the decrease of EYR and increase of ELR over time. Evidently, pushed by the increasing external investment and biogas construction, more efficiency and sustainability of the overall BLAS as well as its subprocesses are needed. An adjustment of the framework of inputs from economy (including NP and RP, both of which are important factors of ESI) should be arranged that is the major concern of LevelIII (next section).

iii. **Resource components tracking (Level III)**

Built on the component-based system decomposition, both the emergy contribution rates (ECRs) and the emergy supply efficiencies (ESEs) of all the resource components within the four subprocesses and overall BLAS were calculated. The

Table 7.18 Emergy accounting of the BLAS in Gongcheng County (2000)

No.	Item	Raw data	Unit	Transformity/ (sej/unit)	References	Solar emergy/ (sej)
1	Solar radiation	3.41E+13	J	1.00E+00	By definition	3.41E+13
2	Rain, chemical	5.28E+10	J	3.06E+04	Odum (1996)	1.61E+15
3	Rain, geopotential	8.41E+10	J	1.49E+04	Odum (1996)	1.26E+15
4	Wind, kinetic	5.14E+11	J	1.11E+03	Odum (1996)	5.72E+14
5	Earth cycle	7.41E+09	J	4.87E+04	Odum (1996)	3.61E+14
Total renewable resources (TRR)						1.97E+15
6	Soil loss	3.35E+09	J	1.24E+05	Brown and Arding (1991)	4.15E+14
Total nonrenewable resources (TNR)						4.15E+14
7	Labor and management	2.37E+01	US $	5.87E+12	Yang et al. (2010)	1.39E+14
8	Seedlings	1.46E-01	US $	5.87E+12	Yang et al. (2010)	8.59E+11
9	Grains	1.32E+08	J	1.39E+05	Lan et al. (2002)	1.84E+13
10	Organic fertilizer	9.97E+06	J	4.54E+06	Odum (1996)	4.52E+13
Total renewable purchases (TRP)						2.04E+14
11	Electricity	1.97E+07	J	3.36E+05	Odum (1996)	6.62E+12
12	Diesels	3.50E+07	J	1.11E+05	Odum (1996)	3.88E+12
13	Machinery	7.42E+03	g	1.12E+10	Odum (1996)	8.31E+13
14	Biogas construction and maintenance	2.81E+00	US $	5.87E+12	Yang et al. (2010)	1.65E+13
15	Pig feed	1.33E+01	US $	5.87E+12	Yang et al. (2010)	7.81E+13
16	Fish feed	6.56E-01	g	5.87E+12	Bastianoni et al. (2001)	3.85E+12
17	Nitrogen	8.79E+04	g	6.38E+09	Odum (1996)	5.61E+14
18	Phosphate	4.05E+04	g	6.55E+09	Odum (1996)	2.65E+14
19	Potash	3.77E+04	g	1.74E+09	Brandt-Williams (2002)	6.57E+13
20	Compound fertilizer	6.39E+04	g	4.70E+09	Lan et al. (2002)	3.00E+14
21	Pesticides	9.35E+03	g	2.52E+10	Brown and Arding (1991)	2.36E+14
22	Infrastructure	9.08E-01	US $	5.87E+12	Yang et al. (2010)	5.33E+12
Total nonrenewable purchases (TNP)						1.63E+15
23	Crops	2.85E+09	J	8.30E+04	Odum (1996)	2.37E+14
24	Fruits	3.94E+09	J	2.70E+04	Odum (1996)	1.06E+14
25	Vegetables	5.82E+08	J	5.30E+04	Odum (1996)	3.08E+13

(continued)

Table 7.18 (continued)

No.	Item	Raw data	Unit	Transformity/ (sej/unit)	References	Solar emergy/ (sej)
26	Teas	2.15E+07	J	2.00E+04	Odum (1996)	4.30E+11
27	Meat product (pig, cow, etc.)	1.35E+09	J	1.71E+06	Odum (1996)	2.32E+15
28	Aquatic product	6.47E+07	J	2.00E+06	Odum (1996)	1.29E+14
29	Biogas	1.27E+09	J	2.48E+05	Bastianoni and Marchettini (2000)	3.16E+14
Total yield (TY)						3.14E+15

numeric results for years 2000, 2004, 2008 were illustrated here (see Tables 7.23, 7.24, 7.25, 7.26, 7.27, and 7.28).

However, the comparative results of ECR and ESE were aggregately shown in Figs. 7.11 and 7.12, addressing the characteristic and dynamic of each subprocess within BLAS. Although herein it is the component-level evaluation that the system is decomposed for, it is actually available for identifying the properties of both the production processes (observed in rows in the tables) and their resource components (observed in columns in the tables) in terms of contribution and efficiency aspects.

Concerning the resource contribution (Tables 7.23, 7.25, and 7.27), the planting production heavily relied on the natural donation (the ECR of both renewable and nonrenewable resources exceed 1), and chemical fertilizer (ECR = 3.44 in 2008) and pesticides have remained the two biggest economic inputs in view of their highest ECRs in all years. Aquaculture production is also highly reliant on natural input (ECR > 1), while labor and management (ECR = 2.36), organic fertilizer (ECR = 1.72), and machinery were among the top three of all the purchased resources. For breeding process, labor and management (ECR > 1) and machinery also contributed a lot, but it was most dependent on pig feed (ECR > 1). Regarding the biogas production, labor and management (ECR = 4.15), machinery (ECR = 1.27) and biogas construction and maintenance (ECR = 1.26) were the top three contributors.

In comparison with ECR, ESE has more violent fluctuations within the subprocesses (Tables 7.24, 7.26, and 7.28). Electricity, diesels, and infrastructure showed the highest supply efficiency (very efficient in supplying the emergy output), far exceeding the average level (ESE = 1) in all processes. Specifically, for planting process, seedlings and grains was also revealed relatively efficient (ESE = 1.01). As to aquaculture process, the supply of fish feed is relatively efficient. And as to the breeding process, labor and management (ESE = 4.92) and pig feed (ESE = 1.88) were efficiently supplied. In biogas production process, machinery and biogas construction and maintenance are efficient factors. The change trends of ECR and ESE over time, roughly speaking, were revealed to be negatively related, which was expectable in the way they were formulated. For

Table 7.19 Decomposed emergy flows within various production processes of the BLAS (year 2000) (*unit* sej)

U (2000)	2	5	6	7	8–9	10	11	12	13	14	15	16	17–20	21	22
Planting	1.48E+15	3.32E+14	3.81E+14	6.99E+13	1.93E+13	3.39E+13	1.99E+12	1.36E+12	3.74E+13	–	–	–	1.19E+15	2.36E+14	2.67E+12
Aquaculture	7.05E+13	1.58E+13	1.82E+13	2.00E+13	–	6.33E+12	1.66E+12	5.82E+11	1.25E+13	–	–	3.85E+12	–	–	5.33E+11
Breeding	3.25E+13	7.29E+12	8.38E+12	1.84E+13	–	4.97E+12	2.32E+12	9.70E+11	1.25E+13	–	7.81E+13	–	–	–	8.00E+11
Biogas	2.71E+13	6.08E+12	6.99E+12	3.07E+13	–	–	6.62E+11	9.70E+11	2.08E+13	1.65E+13	–	–	–	–	1.33E+12
Overall	1.61E+15	3.61E+14	4.15E+14	1.39E+14	1.93E+13	4.52E+13	6.62E+12	3.88E+12	8.31E+13	1.65E+13	7.81E+13	3.85E+12	1.19E+15	2.36E+14	5.33E+12

Table 7.20 Decomposed emergy flows within various production processes of the BLAS (year 2004) (*unit* sej)

U (2004)	2	5	6	7	8–9	10	11	12	13	14	15	16	17–20	21	22
Planting	1.59E+15	3.32E+14	3.66E+14	1.23E+14	2.27E+13	7.13E+13	2.38E+12	1.90E+12	4.04E+13	–	–	–	1.16E+15	2.30E+14	2.58E+12
Aquaculture	7.57E+13	1.58E+13	1.74E+13	3.52E+13	–	1.33E+13	1.99E+12	8.16E+11	1.35E+13	–	–	5.26E+12	–	–	5.15E+11
Breeding	3.49E+13	7.29E+12	8.04E+12	3.25E+13	–	1.05E+13	2.78E+12	1.36E+12	1.35E+13	–	9.99E+13	–	–	–	7.73E+11
Biogas	2.91E+13	6.08E+12	6.70E+12	5.41E+13	–	–	7.94E+11	1.36E+12	2.25E+13	2.08E+13	–	–	–	–	1.29E+12
Overall	1.73E+15	3.61E+14	3.98E+14	2.45E+14	2.27E+13	9.51E+13	7.94E+12	5.44E+12	8.98E+13	2.08E+13	9.99E+13	5.26E+12	1.16E+15	2.30E+14	5.15E+12

Table 7.21 Decomposed emergy flows within various production processes of the BLAS (year 2008) (*unit* sej)

U (2008)	2	5	6	7	8–9	10	11	12	13	14	15	16	17–20	21	22
Planting	1.48E+15	3.32E+14	3.57E+14	1.79E+14	3.15E+13	2.00E+14	2.89E+12	2.13E+12	4.34E+13	–	–	–	1.14E+15	2.07E+14	3.19E+12
Breeding	7.05E+13	1.58E+13	1.70E+13	5.12E+13	–	3.74E+13	2.41E+12	9.12E+11	1.45E+13	–	–	6.43E+12	–	–	6.39E+11
Aquaculture	3.25E+13	7.29E+12	7.84E+12	4.73E+13	–	2.94E+13	3.37E+12	1.52E+12	1.45E+13	–	1.24E+14	–	–	–	9.58E+11
Biogas	2.71E+13	6.08E+12	6.53E+12	7.88E+13	–	–	9.64E+11	1.52E+12	2.41E+13	2.39E+13	–	–	–	–	1.60E+12
Overall	1.61E+15	3.61E+14	3.88E+14	3.56E+14	3.15E+13	2.67E+14	9.64E+12	6.08E+12	9.66E+13	2.39E+13	1.24E+14	6.43E+12	1.14E+15	2.07E+14	6.39E+12

Notes Denotation of numeral items can be found on Table 1. For brevity, items 8–9 and 17–20 are aggregated and evaluated, the same hereafter. Emergy values of items 2, 5, and 6 in different production processes are proportional to the area of different agricultural land use, i.e., 0.92(Planting):0.04(Breeding):0.02(Aquaculture):0.02(Biogas); item 7 is proportional to labor allocation (0.50:0.14:0.13:0.22); item 10 is proportional to the distribution of biogas slurry and residues (0.75:0.14:0.11:0.00); item 12–14 is proportional to the usage intensity of these resources, i.e., (0.30:0.25:0.35:0.10) for electricity, (0.35:0.15:0.25:0.25) for diesels, (0.45:0.15:0.15:0.25) for machinery; item 22 is proportional to the money spent on these production processes, i.e., (0.50:0.10:0.15:0.25); other items not stated are peculiar resources for specific processes

Table 7.22 Decomposed emergy indices within various production processes of the BLAS from 2000 to 2008

Production system	Year	Total Emergy input/(sej)	Total Emergy yield/(sej)	Output efficiency	Transformity (with L&S)/(sej/J)	Transformity (without L&S)/(sej/J)	EYR	ELR	ESI
Planting	2000	3.79E+15	3.75E+14	0.10	5.12E+05	5.02E+05	2.38	1.09	2.18
	2004	3.94E+15	3.96E+14	0.10	5.09E+05	4.93E+05	2.38	0.93	2.57
	2008	3.98E+15	4.32E+14	0.11	4.85E+05	4.63E+05	2.20	1.20	1.83
Aquaculture	2000	1.50E+14	1.29E+14	0.86	2.32E+06	2.20E+04	3.30	0.74	4.48
	2004	1.79E+14	1.36E+14	0.76	2.65E+06	2.13E+06	2.54	1.16	2.20
	2008	2.17E+14	1.58E+14	0.73	2.75E+06	2.10E+06	1.91	1.51	1.26
Breeding	2000	1.66E+14	2.32E+15	13.92	1.23E+05	7.31E+04	1.41	3.18	0.44
	2004	2.11E+14	2.64E+15	12.51	1.37E+05	1.16E+05	1.31	4.59	0.29
	2008	2.68E+14	3.14E+15	11.72	1.46E+05	1.20E+05	1.22	5.74	0.21
Biogas	2000	1.11E+14	3.16E+14	2.85	8.72E+04	2.96E+04	1.57	2.35	0.67
	2004	1.43E+14	3.61E+14	2.52	9.81E+04	4.66E+04	1.42	4.40	0.32
	2008	1.71E+14	4.45E+14	2.60	9.51E+04	3.78E+04	1.30	4.14	0.31
Overall	2000	4.21E+15	3.14E+15	0.75	4.18E+05	1.56E+06	2.30	1.14	2.02
	2004	4.48E+15	3.53E+15	0.79	4.14E+05	3.89E+05	2.25	1.14	1.97
	2008	4.64E+15	4.17E+15	0.90	3.89E+05	3.57E+05	2.04	1.35	1.50

Notes L&S stands for labor and services, i.e., items "7" and "14" in this case

Table 7.23 Emergy contribution rates (ECRs) of different resources within each production process of the BLAS (year 2000)

ECR (2000)	2	5	6	7	8-9	10	11	12	13	14	15	16	17-20	21	22
Planting	4.68	1.05	1.21	0.22	0.06	0.11	0.01	0.00	0.12	–	–	–	3.77	0.75	0.01
Aquaculture	4.70	1.05	1.21	1.33	–	0.42	0.11	0.04	0.83	–	–	0.26	–	–	0.04
Breeding	1.96	0.44	0.51	1.11	–	0.30	0.14	0.06	0.75	–	4.71	–	–	–	0.05
Biogas	2.20	0.49	0.57	2.50	–	–	0.05	0.08	1.69	1.34	–	–	–	–	0.11
Overall	2.71	0.61	0.70	1.03	0.01	0.17	0.06	0.04	0.68	0.27	0.94	0.05	0.75	0.15	0.04

Table 7.24 Emergy supply efficiencies (ESEs) of different resources within each production process of the BLAS (year 2000)

ESE (2000)	2	5	6	7	8–9	10	11	12	13	14	15	16	17–20	21	22
Planting	0.02	0.10	0.09	0.48	1.74	0.99	16.86	24.66	0.90	–	–	–	0.03	0.14	12.56
Aquaculture	0.16	0.73	0.63	0.58	–	1.82	6.96	19.79	0.92	–	–	2.99	–	–	21.61
Breeding	6.37	28.40	24.71	11.24	–	41.66	89.40	213.55	16.62	–	2.65	–	–	–	259.09
Biogas	1.04	4.64	4.04	0.92	–	–	42.62	29.09	1.36	1.71	–	–	–	–	21.17
Overall	1.90	8.47	7.37	3.30	0.43	11.12	38.96	71.77	4.95	0.43	0.66	0.75	0.01	0.04	78.61

Table 7.25 Emergy contribution rates (ECRs) of different resources within each production process of the BLAS (year 2004)

ECR (2004)	2	5	6	7	8–9	10	11	12	13	14	15	16	17–20	21	22
Planting	4.83	1.01	1.11	0.37	0.07	0.22	0.01	0.01	0.12	–	–	–	3.53	0.70	0.01
Aquaculture	4.23	0.88	0.97	1.97	–	0.74	0.11	0.05	0.75	–	–	0.29	–	–	0.03
Breeding	1.66	0.35	0.38	1.54	–	0.50	0.13	0.06	0.64	–	4.73	–	–	–	0.04
Biogas	1.83	0.38	0.42	3.41	–	–	0.05	0.09	1.41	1.31	–	–	–	–	0.08
Overall	3.14	0.65	0.72	1.82	0.02	0.36	0.07	0.05	0.73	0.33	1.18	0.07	0.88	0.17	0.04

Table 7.26 Emergy supply efficiencies (ESEs) of different resources within each production process of the BLAS (year 2004)

ESE (2004)	2	5	6	7	8–9	10	11	12	13	14	15	16	17–20	21	22
Planting	0.02	0.10	0.09	0.27	1.48	0.47	14.09	17.63	0.83	–	–	–	0.03	0.15	13.03
Aquaculture	0.15	0.73	0.66	0.33	–	0.87	5.81	14.12	0.86	–	–	2.19	–	–	22.38
Breeding	6.40	30.68	27.83	6.89	–	21.39	80.51	164.51	16.61	–	2.24	–	–	–	289.62
Biogas	1.05	5.03	4.57	0.57	–	–	38.53	22.50	1.36	1.47	–	–	–	–	23.76
Overall	1.91	9.14	8.29	2.01	0.37	5.68	34.73	54.69	4.91	0.37	0.56	0.55	0.01	0.04	87.20

Table 7.27 Emergy contribution rates (ECRs) of different resources within each production process of the BLAS (year 2008)

ECR (2008)	2	5	6	7	8–9	10	11	12	13	14	15	16	17–20	21	22
Planting	4.46	1.00	1.07	0.54	0.09	0.60	0.01	0.01	0.13	–	–	–	3.44	0.62	0.01
Aquaculture	3.25	0.73	0.78	2.36	–	1.72	0.11	0.04	0.67	–	–	0.30	–	–	0.03
Breeding	1.21	0.27	0.29	1.76	–	1.10	0.13	0.06	0.54	–	4.61	–	–	–	0.04
Biogas	1.43	0.32	0.34	4.15	–	–	0.05	0.08	1.27	1.26	–	–	–	–	0.08
Overall	2.59	0.58	0.62	2.20	0.02	0.86	0.07	0.05	0.65	0.32	1.15	0.07	0.86	0.16	0.04

Table 7.28 Emergy supply efficiencies (ESEs) of different resources within each production process of the BLAS (year 2008)

ESE (2008)	2	5	6	7	8–9	10	11	12	13	14	15	16	17–20	21	22
Planting	0.02	0.10	0.09	0.18	1.01	0.16	11.07	15.04	0.74	–	–	–	0.03	0.15	10.02
Aquaculture	0.17	0.74	0.69	0.23	–	0.31	4.86	12.83	0.81	–	–	1.82	–	–	18.33
Breeding	7.15	31.89	29.67	4.92	–	7.92	68.94	153.04	16.06	–	1.88	–	–	–	242.81
Biogas	1.22	5.42	5.05	0.42	–	–	34.19	21.69	1.37	1.38	–	–	–	–	20.65
Overall	2.14	9.54	8.87	1.44	0.25	2.10	29.76	50.65	4.74	0.34	0.47	0.45	0.01	0.04	72.95

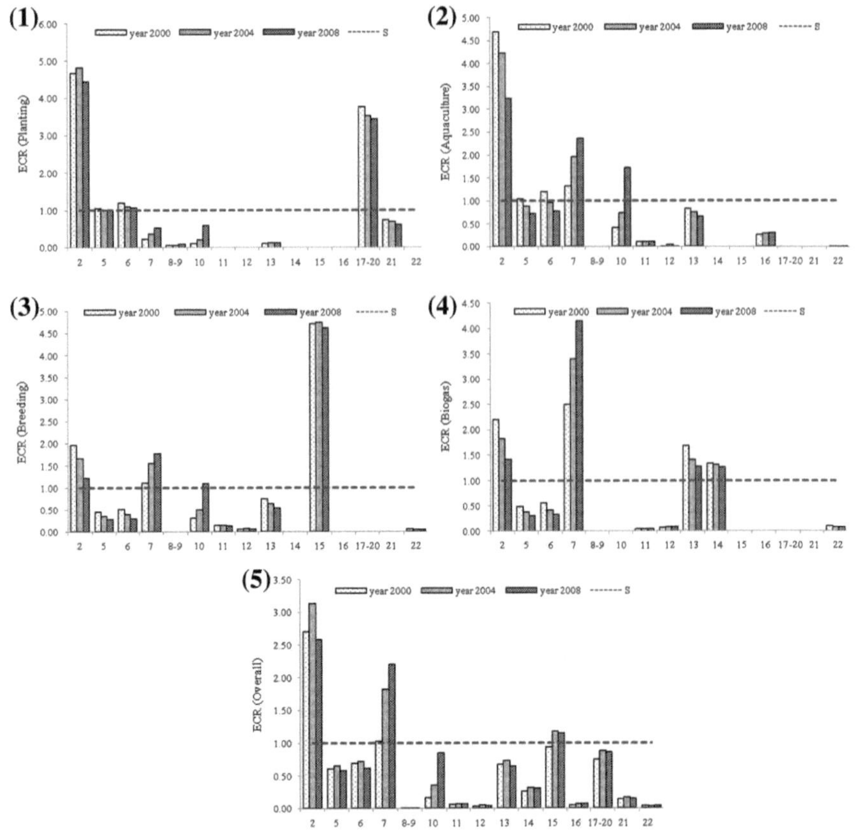

Fig. 7.11 ECRs of different resources with each production process of the BLAS over time: *1* planting production, *2* aquaculture production, *3* breeding production, *4* biogas production, and *5* overall production process. Reprinted from Chen and Chen (2014), with permission from Elsevier

instance, a renewable resource (i.e., labor and management) has been a more contributive part for the planting production process as shown by the increased ECR value, while its efficiency (and capability) of supplying the process has declined according to the lowered ESE (Figs. 7.11a and 7.12a). For the soil loss supplied in the biogas production (Figs. 7.11d and 7.12d), the condition of surface earth erosion has been remedied in this process over time. The supply of this nonrenewable resource has been more efficient which is indicated by the increasing ESE (less erosion is necessary to do the work). However, this opposite tendency of these two indices is not deterministic. In fact, they can be complementary in revealing the dynamic of various processes, together making the ambiguous trends more certain and discernible. In terms of planting production process (Figs. 7.11a and 7.12a), three renewable purchases (i.e., labor and management, seedlings and grains, and organic fertilizer) have been revealed more contributive and less efficient. In contrast, two environmental resource components (earth cycle and soil

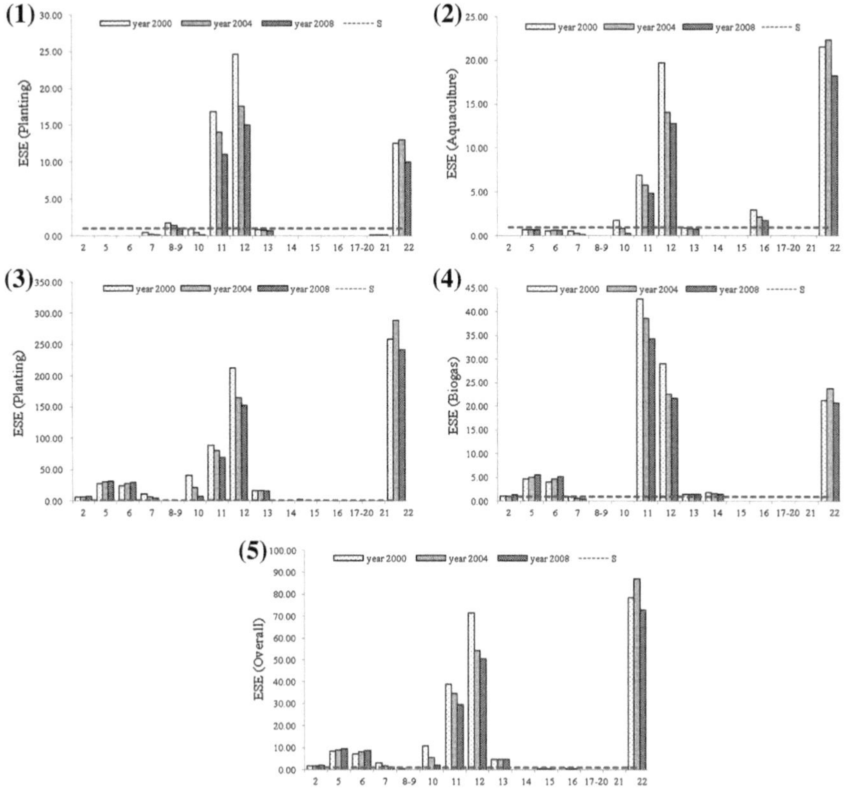

Fig. 7.12 ESEs of different resources within each production process of the BLAS over time: *1* planting production, *2* aquaculture production, *3* breeding production, *4* biogas production overall production process. "S line" stands for average level (which equals 1.00), and *5* overall production process. Reprinted from Chen and Chen (2014), with permission from Elsevier

loss) and two economic inputs (chemical fertilizers and pesticides) were solely observed their increment in ECR, while the dynamics of ESE were uncertain. Again, clear change trends of electricity, diesels, and machinery can only be provided by ESE, showing the continual decrease of their supply efficiencies. As to aquaculture production (Figs. 7.11b and 7.12b), the soil loss has been less extensive and the supply of which has been more efficient. On the contrary, three purchased resources, i.e., labor and management, organic fertilizer, and fish feed, have become more contributive but less efficient since 2000. And the change trend of supply efficiency of diesels to this process has decreased, though the contribution of which is undetermined. The machinery utilized in this process has been surprisingly less contributive and efficient at the same time, in which situation both indices would be essential to articulate the behavior of the resource component. As to breeding process (Figs. 7.11c and 7.12c), the contribution of the free resources from nature has decreased, while the supply efficiency of them has risen. Opposite trend was detected in the utilization of labor and management and organic fertilizer; i.e., they have become more contributive and less intensively used in this process. Clear change tendency for electricity, diesels, and pig feed was only available for ESE, exhibiting an obvious decrement over years. And machinery again has been less contributive and efficient in the process. For biogas production process (Figs. 7.11d and 7.12d), similarly, the contribution of natural resources has decreased, while the supply efficiency of them has gone up. Electricity and diesels have been less efficient due to the decreased ESE, whereas the contribution of machinery has depressed since 2000 according to ECR. Both the contribution and supply efficiency of biogas construction and maintenance have slightly decreased, which turn out to be a problem in front of extensive biogas-related investment. And finally for the overall BLAS (Figs. 7.11e and 7.12e), in terms of emergy contribution, only labor and management and organic fertilizer showed a distinct increase trend among all the resource components. The emergy supply efficiencies of free natural resources have all risen, while those of almost all the economic inputs have displayed a decrement tendency to different extents, which all together produce slightly increasing output efficiency of the whole process as discussed in the Level II evaluation. Our findings suggested that ECR and ESE are two differently focused indices and cannot substitute for each other. Especially, when parsing into the detailed dynamics of process- or component-targeted emergetic evaluation, both indices are critical for innovating current optimization activities toward a better system.

7.3.3 Conclusions

Emergy analysts measure the "real" work of nature as well as human society in the belief of retrospective modeling of existences and their historical courses, on which all the evaluations are based. In essence, the innovation of emergy thinking for system evaluation is that it offers an approach to picture the "presence" (a product or service) through the tracking of "absence" (energy agents) and therefore expands our horizon of renewability and efficiency (Ju and Chen 2011). The explicit inspection into concrete production activities at multiple levels will undoubtedly strengthen this process.

References

Almeida CMVB, Barrell FA, Giannetti BF. Emergetic ternary diagrams: five examples for application in environmental accounting for decision–making. J Clean Prod. 2007;15 (1):63–74.

Bastianoni S, Marchettini N. The problem of co-production in environmental accounting by emergy analysis. Ecol Model. 2000;129(2–3):187–93.

Bastianoni S, Marchettini N, Panzieri M, Tiezzi E. Sustainability assessment of a farm in the Chianti Area (Italy). J Clean Prod. 2001;9:365–73.
Brandt–Williams SL. Folio 4. Emergy of Florida Agriculture. Handbook of Emergy Evaluation: A Compendium of Data for Emergy Computation Issued in a Series of Folios. Gainesville: Center for Environmental Policy, University of Florida, Gainesville, FL, USA; 2002.
Brown MT, Arding, JE. Tranformities working paper. Center for Wetlands, Environmental Engineering Sciences, University of Florida, Gainesville; 1991.
Brown MT, Bardi E. Folio #3: Emergy of ecosystems. Handbook of Emergy Evaluation: A Compendium of Data for Emergy Computation Issued in a Series of Folios. Gainesville: Center for Environmental Policy, Environmental Engineering Sciences, University of Florida. Gainesville, FL, USA; 2001.
Brown MT, Ulgiati S. Emergy–based indices and ratios to evaluate sustainability: monitoring economies and technology toward environmentally sound innovation. Ecol Eng. 1997;9 (1–2):51–69.
Brown MT, Ulgiati S. Emergy evaluations and environmental loading of electricity production systems. J Clean Prod. 2002;10(4):321–34.
Brown MT, Ulgiati S. Energy quality, emergy, and transformity: H.T. Odum's contributions to quantifying and understanding systems. Ecol Model. 2004a;178(1–2):201–13.
Brown MT, Ulgiati S. Emergy analysis and environmental accounting. Encycl Energy. 2004b;2:329–54.
Brown MT, Cohen MJ, Bardi E, Ingwersen WW. Species diversity in the Florida Everglades, USA: a systems approach to calculating biodiversity. Aquat Sci. 2006;68(3):254–77.
Campbell DE. Emergy analysis of human carrying capacity and regional sustainability: an example using the State of Maine. Environ Monit Assess. 1998;51(1):531–69.
Cavalett O, De Queiroz J, Ortega E. Emergy assessment of integrated production systems of grains, pig and fish in small farms in the South Brazil. Ecol Model. 2006;193(3–4):205–24.
Chen SQ, Chen B. Sustainability and future alternatives of biogas-linked agrosystem (BLAS) in China: an emergy synthesis. Renew Sustain Energy Rev. 2012;16(6):3948–59.
Chen SQ, Chen B. Energy efficiency and sustainability of complex biogas systems: a 3-level emergetic evaluation. Appl Energy. 2014;115:151–63.
Chen GQ, Jiang MM, Chen B, Yang ZF, Lin C. Emergy analysis of Chinese agriculture. Agr Ecosyst Environ. 2006;115(1):161–73.
Cheng H. Handbook of natural resource in China. Beijing: Science Press; 1990. p. 545–69 (In Chinese).
Coelho O, Ortega E, Comar V. Balanço de Emergia do Brasil (Dados de 1996, 1989e1981). (Emergy balance of Brazil – Statistics of 1996, 1989 e 1981). In: Engenharia Ecológica Agricultura Sustentável (Ecological Engineering and Sustainable Agriculture); 2003. http://www.fea.unicamp.br/docentes/ortega/livro/index.htm Accessed 15 Jul 2004.
Cohen MJ, Brown MT, Shepherd KD. Estimating the environmental costs of soil erosion at multiple scales in Kenya using emergy synthesis. Agr Ecosyst Environ. 2006;114(2–4):249–69.
Dong XB, Ulgiati S, Yan MC, Zhang XS, Gao WS. Energy and emergy evaluation of bioethanol production from wheat in Henan Province, China. Energy Policy. 2008;36(10):3882–92.
Giampietro M, Ulgiati S. Integrated assessment of large–scale biofuel production. Crit Rev Plant Sci. 2005;24(5–6):365–84.
Giannantoni C, Mirandola A, Tonon S, Ulgiati S. Energy–based, four–sector diagram of benefits as a decision making tool. In: Sergio Ulgiati, editor. Advances in energy studies, vol. 3. Porto Venere, Italy: SGE Editorial; 2002. p. 575–86.
Giannetti BF, Barrell FA, Almeida CMVB. A combined tool for environmental scientists and decision makers: ternary diagrams and emergy accounting. J Clean Prod. 2006;14:201–10.
Hawking S, Mlodinow L. The grand design. New York: Bantam Books; 2010.
He Y, Zhang X, Wen A. Discussion on karst soil erosion mechanism in karst mountain area in southwest China. Ecol Environ Sci. 2009;18:2393–8 [in Chinese].

Jiang MM, Chen B, Zhou JB, Tao FR, Li Z, Yang ZF, Chen GQ. Emergy account for biomass resource exploitation by agriculture in China. Energy Policy. 2007;35(9):4704–19.

Ju LP, Chen B. Embodied energy and emergy evaluation of a typical biodiesel production chain in China. Ecol Model. 2011;222(14):2385–92.

Lan SF, Qin P, Lu HF. Emergy analysis of eco-economic system. Beijing: Chemical Industry Press; 2002 (in Chinese).

Odum HT. Self–organization, transformity, and information. Science. 1988;242(4882):1132–9.

Odum HT. Environmental accounting: emergy and environmental decision making. New York: Wiley; 1996.

Odum HT. Folio 2. Emergy of Global Processes. Handbook of Emergy Evaluation: A Compendium of Data for Emergy Computation Issued in a Series of Folios. Center for Environmental Policy, Environmental Engineering Sciences, University of Florida, Gainesville, FL; 2000.

Odum HT, Odum EC. Energy analysis overview of nations. Working Paper WP–83–82, International Institute for Applied Systems Analysis, Laxenburg, Austria; 1983. p. 421.

Tonon S, Brown MT, Mitandola A, Stoppato A, Ulgiati S. Integration of thermodynamic, economic and environmental parameter for the evaluation of energy systems. In: Sergio Ulgiati, editor. Advances in energy studies, vol. 2. Porto Venere, Italy: SGE Editorial; 2000. p. 635–47.

Ulgiati S, Ascione M, Zucaro A, Campanella L. Emergy-based complexity measures in natural and social systems. Ecol Ind. 2011;11(5):1185–90.

Ulgiati S, Brown MT. Monitoring patterns of sustainability in natural and man–made ecosystems. Ecol Model. 1998;108:23–36.

Ulgiati S, Brown MT. Quantifying the environmental support for dilution and abatement of process emissions: the case of electricity production. J Clean Prod. 2002;10:335–48.

Ulgiati S, Odum HT, Bastianoni S. Emergy analysis of Italian agricultural system. The role of energy quality and environmental inputs. Trends in Ecological Physical Chemistry. Elsevier, Amsterdam; 1993. p. 187–215.

Wang MX, Xia XF, Chai YH, Liu JG. Life cycle energy conservation and emissions reduction benefits of rural household biogas project. Trans Chin Soc Agric Eng. 2010;26:245–50 (in Chinese).

Wei XM, Chen B, Qu YH, Lin C, Chen GQ. Emergy analysis for 'Four in One' peach production system in Beijing. Commun Nonlinear Sci Numer Simul. 2009;14(3):946–58.

Yang J, Chen B. Emergy analysis of a biogas-linked agricultural system in rural China—a case study in Gongcheng Yao Autonomous County. Appl Energy. 2014;118(1):173–82.

Yang ZF, Jiang MM, Chen B, Zhou BJ, Chen GQ, Li SC. Solar emergy evaluation for Chinese economy. Energy Policy. 2010;38:875–86.

Zhang LX, Yang ZF, Chen GQ. Emergy analysis of cropping–grazing system in Inner Mongolia Autonomous Region, China. Energy Policy. 2007a;35(7):3843–55.

Zhang PD, Jia GM, Wang G. Contribution to emission reduction of CO_2 and SO_2 by household biogas construction in rural China. Renew Sustain Energy Rev. 2007b;11:1903–12.

Zhou SY, Zhang B, Cai ZF. Emergy analysis of a farm biogas project in China: a biophysical perspective of agricultural ecological engineering. Commun Nonlinear Sci Numer Simul. 2010;15(5):1408–18.

Chapter 8
Sustainability and Indicator System

Abstract To evaluate the effects of biogas on agro-ecosystem from a systematic perspective, we discussed the current situation of household biogas and identified its main factors that may have impacts on agro-ecosystem. An indicator framework covering social, environmental, and economic aspects was set up. A case study of Gongcheng was then conducted to evaluate the combined impact of biogas project through the proposed indicator framework. In addition, to integrate the socioeconomic-environmental relationships, we innovatively introduced extended exergy analysis to the sustainability evaluation of biogas project. Besides, new extended exergy-based sustainability evaluation indicators that unified greenhouse economic performance, gas emissions, and resource depletion were first proposed to identify biomass conversion pathway planning and as goals for possible system sustainability optimization. The sustainability of a "Three-in-One" biogas production system in southern China was then evaluated based on the proposed framework. Results showed that there was a significant positive effect resulted from the application of biogas, and the integrated benefits have been improved by 60.36%, implying that biogas as a substitute energy source can promote the sustainable level of rural areas. According to the extended exergy analysis, the biogas project has a higher renewability (0.925) and economic return on investment ratio (6.82) and a lower GHG emission intensity (0.012) compared with other renewable energy conversion systems.

Keywords Household biogas · Sustainability · Extended exergy biogas · Agro-ecosystem

8.1 Extended Exergy-Based Sustainability Accounting

The extended exergy analysis, a viable approach for the design and analysis of energy systems using the conservation of mass and conservation of energy principles together with the second law of thermodynamics, is provided, and both the economic properties and environmental characteristics are considered to analyze the

B. Chen et al., *Biogas Systems in China*, DOI 10.1007/978-3-662-55498-2_8

sustainability of household biogas production system. In this chapter, we conducted a detailed cost–benefit analysis according to resource, economic, and greenhouse gas emission for the biogas project. In order to integrate the socioeconomic-environmental relationships, we employed extended exergy analysis to evaluate the sustainability level of biogas project. In addition, resource depletion, greenhouse gas emissions, and economic performance were unified with exergy-based sustainability indicators, which were first proposed to shed light on biomass conversion pathway planning and as goals for biogas system sustainability optimization.

8.1.1 System Boundary

This chapter focused on a unified framework of extended exergy for the sustainability evaluation of the biogas project. The biogas system was assessed by covering the exergy requirements, economic costs, and GHG emissions from the production of resources (cradle) to the end use phase (grave), which included the conversion of the feedstock into biogas (operation), the construction and the comprehensive utilization of biogas and by-products.

The system boundary of the biogas system consists of raw material extraction, processing, transport, and construction of the 8 m^3 biogas digester, biogas production, and multiutilization of biogas and digestate. The operation of the biogas reactor, which is based on manure and straw as feedstock, is also included. The substitution of fossil fuel (by biogas), chemical fertilizer (by residue), the saving of pesticide, and feed (by slurry) are considered in the comprehensive utilization phase.

8.1.2 Results and Discussion

i. Resource depletion

The cumulative exergy inputs required to operate a household biogas digester are summarized in Table 8.1. During a 10-year lifespan, the cumulative exergy demand of a household biogas digester is calculated to be 2.63E+05 MJ, while the exergetic output is 1.83E+05 MJ, which consists of biogas, biogas slurry, and biogas residue. The conversion efficiency is 35.40% if only biogas is used for combustion. When the biogas slurry and residue are fully used, the conversion efficiency can reach 69.39%, indicating the significance of utilization of biogas digestates is to improve resource consumption efficiency.

Various inflows were considered and weighted via multiplying by their respective conversion factors. An exergy signature can be drawn as demonstrated in Fig. 8.1. It can be seen that feedstock (manure and straw) is necessary for biogas production accounts for the largest proportion of total exergy input, followed by the

Table 8.1 Inventory analysis of the cumulative exergy demand of a household biogas digester

Stage	Item	Quantity	Unit	Exergy coefficient	Unit	Reference	Exergy (MJ)
Construction stage	Brick	1320	Piece	1.00	MJ/kg	Qi et al. (2012)	3.28E+03
	Stone	1	m^3	0.32	MJ/kg	Finnveden and Östlund (1977)	8.00E+02
	Sand	3.5	m^3	0.03	MJ/kg	Finnveden and Östlund (1977)	2.97E+02
	Cement	950	kg	4.38	MJ/kg	Finnveden and Östlund (1977)	4.16E+03
	Water	950	kg	0.05	MJ/kg	Dincer and Rosen (2007)	4.75E+01
	Pipe	2	m	32.50	MJ/kg	Chen and Qi (2007)	3.38E+00
	Plastic film	23	m^2	32.50	MJ/kg	Chen and Qi (2007)	1.03E+03
	Steel	8	kg	7.04	MJ/kg	Szargut (2005)	5.63E+01
	Transportation	100	km	1.07	J/J	Kotas (1985)	1.51E+03
Operation stage	Manure	1813	kg/year	12.54	MJ/kg	Qi et al. (2012)	2.27E+04
	Straw	94	kg/year	17.36	MJ/kg	Bösch et al. (2012)	1.63E+03
	Water	5093	kg/year	0.05	MJ/kg	Dincer and Rosen (2007)	2.55E+02
	Leakage	4.5	m^3/year	52.10	MJ/kg	Bösch et al. (2012)	3.02E+02
	Transportation	10	km/year	1.07	J/J	Kotas (1985)	1.51E+02
Utilization stage	Transportation	10	km/year	1.07	J/J	Kotas (1985)	1.51E+02
Cumulative exergy input (10 years)							2.63E+05
Yield	Biogas	450	m^3/year	16.07	MJ/kg	Bösch et al. (2012)	9.33E+03
	Biogas slurry	21,784	kg/year	0.08	MJ/kg	Qi et al. (2012)	1.68E+03
	Biogas residue	3216	kg/year	9.27	MJ/kg	Qi et al. (2012)	2.98E+04
Total yield (10 years)							1.83E+05

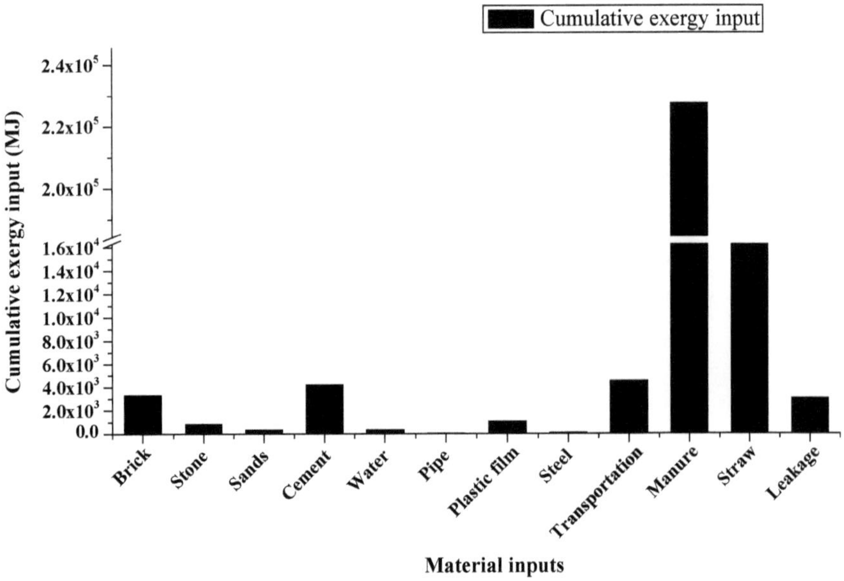

Fig. 8.1 Cumulative exergy inputs of the biogas project

resources exhausted in the construction of the biogas digester, such as brick, cement, and transportation.

ii. Economic performance

Economic costs of the biogas digester occur throughout the lifetime, which can be divided into three stages: construction stage, operation stage, and utilization stage. The cost flows include all expenses for supporting the operation of biogas digester. As an ecological engineering, it is assumed that the biogas project brings about important economic benefits, e.g., biogas use can be substitute for conventional energy consumption and reduce energy expenditures, biogas digestate can be used as organic fertilizer to reduce chemical and pesticide consumption in planting systems, while biogas slurry can be mixed with feed to raise pigs. The economic inputs in the construction and operation stages and the economic benefits in the utilization stage are converted to present values with a discount rate of 8% to investigate economic performance.

Considering the biogas digester construction specifications and biogas production capacity, Table 8.2 summarizes the computed result of the economic costs and benefits of a biogas digester. The total economic costs are calculated to be 4.15E+03 yuan, of which construction cost, operation cost, and utilization cost constitute 52.41, 34.91, and 12.68%, respectively. A household's direct economic benefit for participating in a biogas project is 6.82 times of the investment, 2.83E+04 yuan, indicating that biogas projects are economically profitable and contribute to improving rural life quality.

Table 8.2 Inventory of the economic inputs of a household biogas digester

Stage	Item	Quantity	Unit	Price Per Unit		Economic value (yuan)
Construction stage	Brick	1320	piece	0.35	yuan/piece	−4.62E+02
	Stone	1	m^3	70	yuan/m^3	−7.00E+01
	Sand	3.5	m^3	50	yuan/m^3	−1.75E+02
	Cement	950	kg	400	yuan/t	−3.80E+02
	Water	950	kg	1.50	yuan/m^3	−1.43E+00
	Pipe	2	m	20	yuan/m	−4.00E+01
	Plastic film	23	m^2	0.74	yuan/m^2	−1.70E+01
	Steel	8	kg	3	yuan/kg	−2.40E+01
	Transportation	100	km	8.06	yuan/L	−3.62E+02
	Labor					−5.12E+02
Operation stage	Labor					−8.11E+02
	Water	5093	kg/year	1.50	yuan/m^3	−1.11E+02
	Transportation	10	km/year	8.06	yuan/L	−5.25E+02
Utilization stage	Transportation	10	km/year	8.06	yuan/L	−5.25E+02
	Fuel substitution					1.07E+04
	Chemical fertilizer reduction					1.45E+04
	Pesticide reduction					2.90E+03
	Feed saving					2.56E+02

iii. Greenhouse gas emissions

Figure 8.2 depicts the calculated greenhouse gas emissions from the construction, operation, and utilization phases. In a life-cycle perspective, when the greenhouse gas emissions embodied in building materials are accounted for, the construction stage contributes the largest proportion of greenhouse gas emissions (87.43%), of which bricks and cement are the two largest contributors (Fig. 8.2). In the operation stage, the greenhouse gas emissions can be attributed to leakage and transportation of biogas in the refueling process, and the greenhouse gas emissions are 186.5 kg CO_2-eq, constituting only 8.26% of total emissions. This indicates that biogas is a clean energy and has great potential as a substitute for conventional energy and in greenhouse gas mitigation. In the utilization phase, the transportation of biogas digestate for multiple uses is considered, which occupies the smallest proportion of 4.32%. The largest three emitters of the material inputs of the biogas project are brick, cement, and transportation, which can be targeted for low-carbon management of biogas projects.

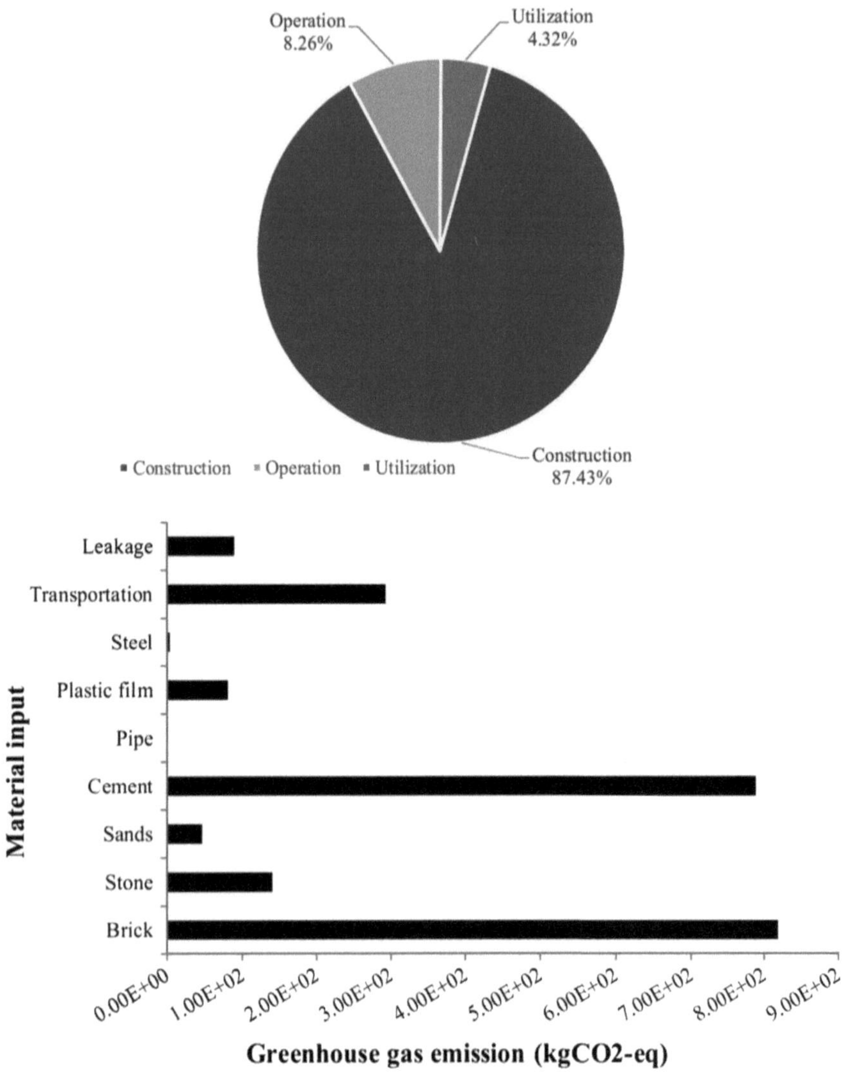

Fig. 8.2 Lifetime greenhouse gas emissions of the biogas project

Biogas as a substitute for conventional fossil fuels can avoid greenhouse gas emissions from coal and direct biomass combustion. The emissions avoided by such use of biogas are calculated, as presented in Table 8.3. The household biogas yield of 450 m^3 can replace 700 kg of coal combustion, increase annual forest protection by 2700 m^3, and save 1500 kg of straw that would have been used for direct combustion, based on the energy structure of the rural areas in South China (Dai et al. 2012). The negative and positive values in Table 8.3 indicate carbon release and reduction, respectively. Assuming that the average forestry storage is

Table 8.3 GHG emission reduction potential of each biogas digester per year

Item	Value	Greenhouse gas emissions or reductions/ton
Substitution of coal	700 kg	+1.74
Substitution of firewood	352 kg	+0.66
Substitution of straw	1500 kg	+2.60
Biogas yield	450 m^3	−1.05
Total	–	+3.95

90 m^3/ha and the wood density is 1.54 kg/m^3 (State Forestry Administration of China 2010), biogas is carbon-reducing, with 3.95 t CO_2 being avoided per household. From the life-cycle perspective, the carbon emissions reduced by biogas utilization over 10 years are 1.74 times the greenhouse gas emissions produced by the biogas project, indicating that biogas utilization plays an important role in the mitigation of greenhouse gas concentrations.

iv. Sustainability indicators

In Table 8.4, the extended exergy-based sustainability indicators that reflect the sustainable level of the biogas project are compared with those of other renewable energy systems.

The ε_P of the biogas project is 0.69 MJ/MJ. Thus, it can be argued that 31% of exergy contents is exhausted within the biogas production process. To maximize the conversion efficiency of the biogas project, it is necessary to adjust fermentation temperatures and feedstock configurations to enhance routine maintenance. The ε_P of the biogas project is similar to that of biodiesel production but much higher than that of a wind farm, compared with other renewable energy systems, indicating that the biogas system has a relatively high technical efficiency.

The *R*% of the biogas project is higher than the other two biodiesel production systems, reaching 92.50%. This high R% occurs because although some nonrenewable exergy inputs, such as construction materials and energy, flow into the biogas production system, these nonrenewable exergy inputs account for only 7.50%, illustrating that the biogas project is competitive in the context of renewability.

Table 8.4 Comparison of different renewable energy utilization projects with extended exergy-based sustainability indicators

Indicators	Biogas project	Wind farm[a]	Biodiesel from used cooking oil[b]	Biodiesel from rapeseed crops[b]
ε_P	0.69	0.028	0.751	0.516
R%	0.925	0.973	0.777	0.529
ERIO	6.820	2.800	1.380	5.824
CI	0.012	0.023	0.006	0.062
SI	568.333	121.739	247.660	94.203

[a]Yang et al. (2012); [b]Talens Peiró et al. (2010)

The *EROI* is used to evaluate the economic efficiency of the biogas project. As presented in Table 8.4, the *EROI* of the biogas project (6.82) is much higher than other renewable energy generation systems, for instance, wind power and biodiesel generated from cooking oil and rapeseed crops. Obviously, biogas production creates prominent economic benefits via the comprehensive utilization of biogas and its by-products.

CI can be a measurement of global warming stress exerted by biogas production. The *CI* of the biogas project is lower than those of the wind power generation system and biodiesel generation from used rapeseed crops, but higher than *CI* of the biodiesel extracted from used cooking oil, which indicates that the biogas project is a promising renewable option to alleviate global warming stress from conventional fossil fuels consumption.

The *SI* is determined by both *EROI* and *CI*. Due to the higher *EROI* and lower *CI* of the biogas project, the *SI* value is higher than those of the wind power generation system and the biodiesel production systems. In addition, by improving the economic conversion efficiency as well as alleviating the environmental costs, better performances are expected from the biogas project.

8.1.3 Conclusions

In this chapter, extended exergy was first introduced as a benchmark for evaluating the sustainability of the biogas production system by providing a unified measurement for material and energy, economic, and greenhouse gas flows in different stages of the lifetime of biogas digester. As an extension of current extended exergy analysis, indicators were also proposed to reflect systematic performance. These unitary thermodynamic indicators may be useful for comparisons of sustainability levels between biogas projects and other renewable utilization pathways as is more eco-oriented, with a significant focus on explaining the relationships among exergy, environmental impacts, and economic features. In addition, the extended exergy-based multicriteria sustainability assessment approach presented can also serve as a vital tool for identification of biogas management hotspots and interventions in policy making.

The main material inputs are manure and straw for a biogas project (Fig. 8.1) so that the biogas project has a high renewability. In addition, it is effective in the transformation of wastes (straw, manure) into useful substances (fuel, feedstock, and fertilizer). Routine maintenance should be enhanced to further improve the conversion efficiency. Except for adjusting the C:N ratio in the fermentation pool, fermentation temperature, and PH value, which are the key factors affecting the biogas productivity, pretreatment of feedstock (e.g., straw pulverization), and stirring in fermentation are effective measures to enlarge the contact surface of microorganism and feedstock, accelerated fermentation rate, thereby increasing gas yield.

The biogas production process is calculated to be carbon-reducing and competitive compared with other biomass utility methods. Through the biogas alternative, a desired level of abatement of greenhouse gas emissions can be achieved (Table 8.3). Therefore, the promotion of biogas in rural areas is in line with the national objectives of "low-carbon society" and "sustainable rural development." Meanwhile, the reductions in greenhouse gas emissions generate the finance for the project, such as through conducting CDM programs, the Certified Emission Reductions (CERs) can be sold on the emissions trading market. Therefore, the high biogas maintenance cost can be compensated by gaining additional revenue. The government should also take measures for biogas projects, such as tax incentives to support the development of CDM programs.

However, according to the results of economic analysis, the large investment in biogas digester construction may preclude the enthusiasm of farmers in promoting biogas projects, although the biogas project is economically profitable with a high economic return on investment ratio. Considering the main material, cement and brick for biogas digester production, the current prices are more than twice of these of 2004, while government subsidy has barely changed. The sharp rise in economic cost aggravated the burden on poor families. Therefore, government subsidy on biogas digesters should be further enhanced. Meanwhile, an incentive mechanism can be established to attract more investments on biogas project so as to simulate the industrialization of biogas project.

Another approach that can be applied to motivate enthusiasm of farmers on the utilization of biogas is the multiple uses of biogas and its co-products. In addition to decreasing fuel costs by using biogas as a substitute for conventional fossil fuels, it should also promote the utilization of biogas residue as organic fertilizer and of biogas slurry as breeding feedstock, a seed soaking material, and a pesticide substitute.

The development of biogas projects in many areas was hindered by the backward infrastructure and service system. Because the subsidy is directly granted for households, no expenditure is set aside for the service station, and the technical guidance and training are precluded. The consequence is, management and service cannot catch up with the construction of biogas digesters, and the biogas digesters cannot be properly maintained. As a result, many biogas digesters are abandoned, and the biogas projects are unsustainable. Therefore, the government should pay more attention to the subsidy of service stations. For example, if the subsidy is improved from 1500 to 2050 yuan/household, 10% of the subsidy will be allocated to the construction of infrastructure and service system. Moreover, with the implementation of "Building Materials to the Countryside" policy launched by the China Building Materials Federation of Industry and Information Technology Ministry, the costs of biogas projects will be further lowered, which is subsided by the government and aims to encourage the construction of energy-saving projects and housing in rural China.

8.2 Indicator System for Impact Assessment

Biogas as a linkage of agro-ecosystem is of great importance in propelling the sustainable development of rural areas, which can reduce the traditional domestic consumption of firewood and coal and thereby alleviate the environmental pressure. As the substitute of chemical fertilizer, biogas slurry and residue could promote the development of cropping by improving soil quality. Meanwhile, social and economic benefits can be obtained through the biogas projects, e.g., improvement of rural sanitation and incomes of farmers.

However, the multifaceted impacts of biogas projects in rural areas ought to be quantified by selecting a range of specific indicators. The application of biogas has significant influences on the agro-ecosystem from the aspects of resource, environment, technology, economy, and society. Since the biogas project is dedicated to promoting the sustainable development of agro-ecosystem, the indicators evaluating the impact of biogas project on agro-ecosystem should give priority to the trade-offs between its economic, social, and ecological objectives that reach acceptable values for the society as a whole (Hediger 1999; Stoorvogel et al. 2004).

An indicator system including environmental, economic, and social factors is established to reflect the influence of biogas on sustainable development through questionnaire investigation and calculation. Using the 5-point grade system and analytic hierarchy process (AHP) method, we quantified the overall score and categorized it into the corresponding grade to assess the compound effects of biogas projects.

8.2.1 Data Sources

The basic data collection, status investigation, and consultation for experienced experts in biogas production are conducted to collect the basic information. The value of each indicator is thereby based on the virtue of questionnaires and thermoeconomic calculations. Environmental impacts act as a changing rate, which is derived from comparing the performance under the circumstance that 90% of the households have turned to use biogas as a major energy source in 2009 with that of nonbiogas application period in 1983, as listed from Eq. (8.1) to (8.6), while the values of social and economic indicators are derived directly from the questionnaires by comparison between biogas households and nonbiogas households.

(1) The impact on soil quality caused by the increase in organic fertilizer is depicted as below:

$$a = G_{\mathrm{or}}/F = \left[N_{\mathrm{t}} \times \left(g_{\mathrm{residue}} \times \alpha_{\mathrm{residue}} \times \theta_{\mathrm{residue}} + g_{\mathrm{slurry}} \times \alpha_{\mathrm{slurry}} \times \theta_{\mathrm{slurry}} + g_{\mathrm{straw}} \times \alpha_{\mathrm{straw}} \times \theta_{\mathrm{straw}}\right)/S\right]/F \quad (8.1)$$

where a is the proportion of the chemical fertilizer substituted by organic fertilizer; G_{or} represents the organic matter increased in the soil; F is the background value of soil; S is the area of arable land; N_t is the number of biogas digesters that are under operation; $g_{residue}$, g_{slurry}, g_{straw} are the residue, slurry produced, and straw saved by biogas production per household, respectively; $a_{residue}$, a_{slurry}, a_{straw} stand for the proportion of residue, slurry, and straw that are put into the farmland for fertilizing the soil, respectively; $\theta_{residue}$, θ_{slurry}, θ_{straw} are the organic matter existed in residue, slurry, and straw, here we use 33.2, 1.29, and 12%, respectively.

(2) CO_2 emissions caused by biomass burning and coal combustion are demonstrated in Eqs. (8.2)–(8.4):

$$C_W = N_t \times W \times 45\% \times 87\% \times 44/12 = 1.436 N_t \times W \tag{8.2}$$

$$C_S = N_t \times S \times 40\% \times 85\% \times 44/12 = 1.247 N_t \times S \tag{8.3}$$

$$C_C = N_t \times C \times 0.0209 \times 24.26 \times 80\% \times 44/12 = 1.487 N_t \times C \tag{8.4}$$

where C_W, C_S, and C_C are CO_2 emissions released by the combustion of firewood, straw, and coal, respectively; W, S, and C represent the firewood, straw, and coal saved per household caused by the substitution of biogas with the unit of t, respectively; 45%, 40% are the coefficients of carbon contained in firewood and straw, respectively; 87, 85, and 80% are the proportion of carbon oxidized firewood, straw, and coal, respectively; 0.029 is the calorific value and 24.26 is the carbon emission factor of coal with the unit of TJ/t and t/TJ, respectively.

Thus, the proportion of reduced CO_2 emission is represented as below:

$$c = (C_W + C_S + C_C)/C_0 \tag{8.5}$$

where C_0 is the CO_2 emission before the use of biogas.

(3) The quantity of SO_2 released currently is calculated and compared with the period of nonbiogas use, as Eqs. (8.6)–(8.7) shown:

$$S_C = N_t \times C \times 16 \times 84\%/1000 = 0.0134C \tag{8.6}$$

$$e = S_C/S_0 \tag{8.7}$$

where S_C is the decreased SO_2 emission; S_0 is the SO_2 emission during the period of nonbiogas.

(4) The reduction of deforestation rate can be calculated in Eq. (8.8).

$$f = W \times N_t/W_0 \tag{8.8}$$

where f is the reduction rate of forest deforestation; W_0 stands for the firewood consumed per year before the use of biogas.

Table 8.5 Five-grade classification

Grades	Rates		Value
Significant beneficial impact	Increase by 90%		9
Large beneficial impact	Increase by 70%		7
Moderate beneficial impact	Increase by 50%		5
Slight beneficial impact	Increase by 30%		3
Minor beneficial impact	Increase by 10%		1
Significant adverse impact	Decrease by 90%		−9
Large adverse impact	Decrease by 70%		−7
Moderate adverse impact	Decrease by 50%		−5
Slight adverse impact	Decrease by 30%		−3
Minor adverse	impact	Decrease by 10%	−1

(5) The proportion of harden land due to biogas digester construction is calculated using the total construction area of biogas digester divided by the total area of arable land.
(6) The proportion of disposed excreta is used to illustrate the impact on hygienic conditions of rural area.

According to the influence level caused by household biogas projects, all these rates are normalized to a five-grade classification, as shown in Table 8.5. The impacts of biogas on different aspects are summarized into two categories covering both positive and passive impacts. A detailed classification is made to further distinguish the different influencing levels. A five-grade classification is finally determined, which includes the grades of severe impact, big impact, moderate impact, slight impact, and minor impact.

8.2.2 *Analytic Hierarchic Process*

AHP has been one of the most widely used multiple criteria decision-making tools for decision makers and researchers (Vaidya and Kumar 2006). Here, the tool is applied to evaluate the impacts of biogas on agro-ecosystem. Based on the framework established in Table 8.4, the important levels of factors that are in the same layer are determined by pairwise comparisons. According to the scale set made by Saaty (1992), we get the weight of each element by comparing these pairs. Finally, the weight of each indicator to the destination layer is obtained layer by layer. In this step, to evaluate the impacts of biogas on sustainable development, we dispatch the comparison tables to the local decision makers and professional researchers.

8.2.3 Results and Discussion

We collected the questionnaires, identified the weight of each indicator, and confirmed the weight through consistency check. Then, the results of comprehensive evaluation are obtained, as listed in Table 8.6.

Table 8.6 Impacts of biogas project on the sustainable development of agro-ecosystem

Destination layer	Criteria layer	Indicator layer	Total weight	Value of indicators	Value of impacts
	Environmental indicators 0.63	Increasing rate of organic matter	0.035	2.49	0.084
		SO_2 reduction rate 0.082	0.052	3.51	0.156
		CO_2 reduction rate 0.385	0.243	5.12	2.187
		Forest deterioration reduction rate 0.177	0.112	8.55	1.008
		Proportion of arable land occupied by biogas digester 0.029	0.018	0.54	−0.018
		Animal excreta disposal rate 0.272	0.171	7.8	1.197
Sustainable development	Social indicators 0.151	Reduction in the rate of emigrate workers 0.067	0.01	2	0.02
		Increase in labor productivity rate in rural areas 0.115	0.017	1.5	0.034
		Increase in the proportion of technical farmers 0.439	0.066	3.1	0.198
		Decrease in infection rate of parasitic disease 0.189	0.029	5.6	0.1624
		Decrease in infection rate of intestine disease 0.189	0.029	9.1	0.261
	Economic indicators 0.218	Promotion in farmers' consumption level 0.059	0.013	8.5	0.104
		Increasing rate of grain output per hectare 0.282	0.06	7.2	0.42
		Increasing rate of pig breeding number 0.176	0.038	1.5	0.076
		Increasing rate of fruit output per hectare 0.443	0.097	5.5	0.485
		Proportion of construction cost 0.039	0.008	1.7	−0.016
Sum					6.3584

As mentioned above, the complex influence is 5.44, which is approximate to moderate beneficial impact according to Table 8.5, implying that biogas as a substitute energy source is beneficial to promote the sustainable development of agro-ecosystem. Although some passive impacts have been brought by the construction of biogas digesters, they can be ignored, e.g., the cost of biogas digester construction is -1.7, which is between minor and slight adverse impact. Therefore, the future development of biogas should be encouraged in order to increase the current moderate beneficial impact on agro-ecosystem. Such results are convincing in spreading the biogas linked-ecological agriculture mode of Gongcheng to the other rural areas of China.

The impacts of the biogas project on environmental, social, and economic subsystems are represented in Table 8.7. It can be found that the environmental impact of biogas is 5.86, which contributes most to the total benefit in Gongcheng. i.e., the application of biogas has produced large environmental and ecological benefits, especially its significant role in decreasing the deforestation area, increasing animal excreta disposal rate, and reducing CO_2 emission, as shown in Table 8.6. The values of impacts on these three aspects are 8.55, 7.8, and 5.12, respectively. Thus, biogas system is significant for sustainable development of agro-ecosystem and the conservation of forests and construction of "low-carbon village" as well.

The biogas can increase the agricultural yield in many ways, particularly the promotion in economic benefits. Its influence on economic subsystem is calculated to be 4.9, which is at a moderate level. Since biogas slurry and residue serving as organic fertilizer can substitute the chemical fertilizer and increase farmers' revenues, the impacts on consumption level and food production are notable, with the value of 8.47 and 7.2, respectively. However, the positive influence on breeding industry in Gongcheng is quite weak. The passive impact brought by the cost of biogas digester construction is relatively small with the value of -1.7. With the help of government subsidy, the obstacle that farmers cannot afford the construction cost will be gradually overcome.

The value of social benefit is quantified to be 4.4, which is between the slight and moderate levels. The most significant influence on social aspect is the decrease in infection rate of intestine disease because of the increase rate of excrements disposal. Moreover, the application of biogas in rural areas can not only provide more employment opportunities and reduce emigration workers, but also make use of the rural surplus labor. Although the ratio of skilled farmer has increased by 30%, trainings still need to be enhanced to cultivate "New Farmers."

Table 8.7 Environmental, social, and economic benefits derived from biogas project

	Environmental benefit	Social benefit	Economic benefit
Score	7.3	4.4	4.9
level	Large beneficial impact	Moderate beneficial impact	Moderate beneficial impact

8.2.4 Conclusions

With the application of the biogas, a moderate beneficial impact is exerted on agro-ecosystem of Gongcheng, in which environmental factors contribute the most. For the environmental subsystem, the use of biogas has produced large environmental and ecological benefits, e.g., decreasing the deforestation area, increasing animal excreta disposal rate, and reducing CO_2 emission. The utilization of biogas and its co-products also promotes the economic subsystem, in which the impacts on consumption level and food production are notable. However, the influence on breeding industry is relatively small in Gongcheng. Moreover, biogas project in rural area has either increased employment opportunities or reduced the incidence of diseases, through which improves the social conditions in Gongcheng.

References

Bösch P, Modarresi A, Friedl A. Comparison of combined ethanol and biogas polygeneration facilities using exergy analysis. Appl Therm Eng. 2012;37:19–29.

Chen GQ, Qi ZH. Systems account of societal exergy utilization: China 2003. Ecol Model. 2007;208(2–4):102–18.

Dai J, Chen B, Qi J. Ecological and economic benefit accounting for low carbon biogas project construction–a case study of Gongcheng, Guangxi. China Popul Res Environ. 2012;22(3):157–63.

Dincer I, Rosen MA. Exergy, environment and sustainable development. Netherland: Elsevier Ltd; 2007. p. 36–59.

Finnveden G, Östlund P. Exergies of natural resources in life–cycle assessment and other applications. Energy. 1997;22(9):923–31.

Hediger W. Reconciling 'weak' and 'strong' sustainability. Int J Soc Econ. 1999; 26 (7/8/9): 1120–1143.

Kotas TJ. The exergy method of thermal plant analysis. Stoneham Ma: Butterworth Publishers; 1985. p. 288–292.

Peiró LT, Méndez GV, Sciubba E, i Durany XG. Extended exergy accounting applied to biodiesel production. Energy 2010; 35(7): 2861–9.

Qi J, Chen B, Chen WC, Chu XL. Inventory analysis for a household biogas system. Procedia Environ Sci. 2012;13(10):1902–6.

Saaty TL. Decision Making for Leaders. Pittsburgh, USA: RWS Publications; 1992.

State Forestry Administration of China. The forestry action plan to address climate change. Beijing: China Forestry Press; 2010.

Stoorvogel JJ, Antle JM, Crissman CC, Bowen W. The trade-off analysis model: integrated bio-physical and economic modeling of agricultural production systems. Agric Syst. 2004;80 (1):43–66.

Szargut J. Exergy method: technical and ecological applications. J Res Pharm Econ. 2005;10 (1):51–81.

Vaidya OS, Kumar S. Analytic hierarchy process: an overview of applications. Eur J Oper Res. 2006;169:1–29.

Yang J, Chen B, Sciubba E. Exergy–based sustainability evaluation of a wind power generation system. In: Proceedings of the 25th International conference on efficiency, cost, optimization and simulation of energy conversion systems and processes; Aabo Akademi University, ECOS 2012. vol 2, p. 263–70.

MIX
Papier aus verantwortungsvollen Quellen
Paper from responsible sources
FSC® C105338

If you have any concerns about our products,
you can contact us on
ProductSafety@springernature.com

In case Publisher is established outside the EU,
the EU authorized representative is:
Springer Nature Customer Service Center GmbH
Europaplatz 3, 69115 Heidelberg, Germany

Printed by Libri Plureos GmbH
in Hamburg, Germany